Unit 3 — Developing Scientific Skills

Unit 4 — Using Scientific Skills for the Benefit of Society

Published by Coordination Group Publications Ltd.

Editors:
Ellen Bowness, Tom Cain, Gemma Hallam, Sarah Hilton,
Andy Park, Rose Parkin, Claire Thompson, Julie Wakeling.

Contributors:
Mike Bossart, James Foster, Vinette Jackson, John Myers, Philip Rushworth, Adrian Schmit,
Claire Stebbing, Moira Steven, Pat Szczesniak, Mike Thompson, Sophie Watkins.

ISBN: 978 1 84146 774 0

With thanks to Barrie Crowther, Mark A. Edwards,
James Foster, Sue Hocking and Glenn Rogers for the proofreading.
With thanks to Katie Steele for the copyright research.

With thanks to Science Photo Library for permission to reproduce the photographs used
on pages 8, 103 and 136.

Data used to construct stopping distance diagram on page 97 from The Highway Code.
Reproduced under the terms of Click-Use license.

Groovy website: www.cgpbooks.co.uk

Printed by Elanders Hindson Ltd, Newcastle upon Tyne.
Jolly bits of clipart from CorelDRAW®

Based on the classic CGP style created by Richard Parsons.

Organisations that Use Science

Science contributes to your <u>everyday life</u> in loads of ways, most of which you probably haven't even thought about — the <u>shampoo</u> you use, the plastic used to make your <u>lunchbox</u>, and even the <u>water</u> you drink are all there because of science organisations...

Organisations that Use Science Can Benefit Society

There are <u>loads</u> of different science organisations and businesses — some <u>make products</u> and some <u>provide a service</u>. Here are some examples of <u>how</u> those products or services benefit society (i.e. you):

By Making Useful Products...

1) DRUGS

<u>Pharmaceutical companies</u> develop, test and manufacture <u>drugs</u> — drugs help us to <u>fight off disease</u>.

2) CHEMICALS

<u>Chemical companies</u> develop and produce things like <u>fertilisers</u> and <u>paint</u> — without these food would be more expensive (and art lessons would be a lot less interesting).

3) FOOD

<u>Food manufacturers</u> grow and process <u>food</u> (unsurprisingly) — they produce large amounts of safe food.

...And Providing Useful Services

1) KEEPING US HEALTHY

The <u>health service</u> includes hospitals, doctors, dental surgeries and pharmacists.

3) ANALYSING CHEMICALS

Lots of organisations have <u>laboratories</u> that do things like make sure the <u>water is safe</u> to drink and make sure <u>products</u> are of a <u>consistent quality</u>.

2) GLOBAL COMMUNICATION

<u>Telecommunications</u> companies make it possible to phone your mate in Australia.

4) EDUCATING PEOPLE

Schools, colleges and universities teach science (lucky for you).

5) PROVIDING ENERGY

Some companies <u>generate</u> and <u>distribute</u> energy — without electricity it'd be hard to do lots of things, let alone watch Neighbours.

These organisations also provide <u>employment</u> for millions of people and large companies <u>generate</u> lots of money for the <u>country</u>. Local organisations are just as important for the <u>local economy</u>.

Organisations may be Local, National or International

1) <u>International</u> organisations have sites in <u>more than one country</u>. They're usually big companies — some examples of those that use science are BP p.l.c., Unilever p.l.c. and GlaxoSmithKline.

2) <u>National</u> organisations are based in just <u>one country</u>, and they distribute their goods or services throughout the country. In the UK there are organisations like the NHS and the Environment Agency, as well as nationwide shop chains. It's sometimes <u>hard</u> to tell whether a company is national or international, but you should be able to find this out from their website (if they have one).

3) <u>Local</u> organisations that use science include things like schools, colleges, health centres and dentists.

Science organisations — even better than sliced bread...

So there you have it — without a supply of people with <u>science training</u> and <u>skills</u> there would be no <u>science organisations</u>. That wouldn't leave us with very much. So, unless you want to live in a really <u>rubbish future</u> (with no cars, TV or even food), you'd better get <u>learning</u> — your country needs you.

Locating an Organisation

Ever wondered <u>why</u> an organisation is based <u>where</u> it is? Believe it or not, a lot of <u>thought</u> goes into where they're <u>located</u> and it's not just things like, "because it's round the corner from Aunty Flo's"...

There are Lots of Factors to Consider

Not all of these factors are relevant for all organisations, but the <u>general</u> reasons behind location are:

Raw materials

The presence of <u>raw materials</u> required for the process.

EXAMPLE: Breweries are often located next to supplies of <u>pure spring water</u> that is essential for the production of good quality beers and spirits.

Workforce

The availability of a <u>workforce</u> with the <u>right skills</u>.

EXAMPLE: Many high-tech companies (such as biotechnology companies) are based on <u>University science parks</u> because it's easy to recruit employees from the University.

Land

The <u>cost</u> of <u>land</u>.

EXAMPLE: The cost of land in the South East is more <u>expensive</u> than elsewhere — many companies have <u>relocated</u> to the Midlands and up North.

Energy

The availability of an <u>energy supply</u>.

EXAMPLE: Aluminium production is sited in Conwy (in Wales) because <u>hydroelectric power</u> can be readily produced there.

Transport links

Good <u>transport links</u> for delivery of raw materials.

EXAMPLE: Oil refineries are located around <u>ports</u> for supplies of crude oil from <u>tankers</u>.

Market

A <u>market</u> for the <u>product</u> or <u>service</u>.

EXAMPLE: Companies who manufacture dyes tend to be found in areas of textile manufacture.

Grants

Availability of <u>Government</u> or <u>European grants</u> to reduce the <u>start-up costs</u>.

There Could be Effects on the Local Environment

The previous page described some of the ways society benefits from organisations that use science. But because of the type of work they do, some organisations can have a <u>damaging effect</u> on the <u>environment</u>.

1) <u>Toxic pollution</u> — nasty chemicals, e.g. from a chemical works, could contaminate the environment.

2) <u>Visual pollution</u> — some factories and company <u>buildings</u> can be pretty <u>unsightly</u>. Also, things like chemical works and oil refineries are sometimes <u>illuminated</u> at night and are a source of <u>light pollution</u>.

3) <u>Noise pollution</u> — big trucks and big machinery are usually noisy. This can be a big problem for locals if the business operates 24 hours a day.

4) <u>Traffic congestion</u> — large businesses, e.g. a brewery, may need <u>frequent deliveries</u> of raw materials and <u>collection</u> of products — if this is done by road the lorries might cause <u>traffic congestion</u> in the local area and <u>damage</u> to road surfaces.

So you wouldn't put a tidal power station in the Sahara...

You may well be wondering <u>why</u> you need to know all this, well — soon you'll have to produce a <u>report</u> into an organisation, explaining the <u>reasons for its location</u> and its <u>effects on the local environment</u>.

Roles of Scientists

Of the UK's workforce, a massive four million people carry out jobs that use science.
You might be surprised how many different jobs there are in science...

Science Qualifications Offer a Range of Different Careers

This page covers just some of the areas in which scientific skills can be used. There are loads of
employment opportunities for people with scientific skills. There just isn't the space to list them all.

HEALTHCARE — e.g. doctors, dentists, nurses, pharmacists, radiographers, physiotherapists. There are also people who support these roles, e.g. medical physicists and lab technicians.

EDUCATION — e.g. secondary school science teachers, university and college lecturers.

ENGINEERING (the development of materials and technology) — e.g. chemical engineers develop paints and dyes, mechanical engineers develop machines.

SCIENCE QUALIFICATION

PHARMACEUTICALS — e.g. research scientists develop, make and test drugs.

FOOD INDUSTRY — e.g. food scientists develop foods for supermarkets and food manufacturers, microbiologists test food to make sure it's safe.

MANUFACTURING — e.g. analytical scientists are involved in quality control (making sure manufactured goods are of a consistent quality).

AGRICULTURE — e.g. research scientists look at new ways to produce foods or monitor standards of production, vets keep animals healthy.

There are loads of others — some scientists work for the police as forensic scientists, there are science editors, scientific patent lawyers, technicians who support the work of other scientists, goat breeders, orangutan urine collectors, and many, many more.

There are Major, Significant or Small Users of Science

1) Major users of science are people who use scientific skills as a large part of their job. They generally have a science-based qualification (see next page), e.g. science teachers, laboratory and research scientists, doctors and nurses.

2) Significant users of science are people who use scientific skills as part of their jobs — their training will have involved learning a fair amount of scientific knowledge, e.g. science editors and many of the healthcare professionals (such as dieticians).

3) Small users of science are people who use basic scientific skills as part of their jobs. They don't work in a science-based job and their training probably wouldn't have included that much science. Small users include hairdressers (carrying out allergy tests before applying hair dyes and bleaches), photographers (using chemical solutions to develop photographs), plumbers and electricians.

This is really just a guide though. There are no hard-and-fast rules about putting people into categories — different people might have different ideas about where they belong.

Rolls of scientists — how does that work then?

Hopefully now you can see that the possibilities really are endless. You could become an engineer, or even find yourself working as an editor for a company that makes revision guides, sharing your wealth of scientific knowledge with the young scientists of tomorrow. Still, don't expect that to make your mam happy — she'll still want you to join the navy.

Skills and Qualifications

Science is a <u>compulsory</u> subject in UK schools until age 16 — after that you can do <u>what you want</u>. Some (usually crazy) people decide that they haven't quite had enough and do even more science after the age of 16. There are loads of different options out there — <u>apprenticeships</u>, <u>degrees</u>, <u>NVQs</u>, the list goes on...

People Who Use Science Usually Have Special Qualifications...

1) People who are <u>major</u> scientific users will usually have a <u>degree</u> in science. This could be a <u>general degree</u>, e.g. in biology or chemistry, or a <u>specialised degree</u>, for example in forensic science or food science.

2) Some scientists, particularly those working in <u>research</u>, will have a <u>higher degree</u> — this can be either a masters (e.g. an MSc) or a PhD (scientists who have done a PhD are then called doctors).

3) For many careers you have to obtain special <u>professional qualifications</u>. Teachers have to do a <u>PGCE</u> (Postgraduate Certificate of Education), which is a special teaching qualification. People working in the <u>healthcare sector</u> have qualifications that test their understanding of science. They also may have to be '<u>registered</u>' with a <u>supervisory body</u> in order to practise.

4) Many organisations run their own <u>training schemes</u> (often linked to <u>Modern Apprenticeships</u> or <u>NVQs</u>) for careers like technicians and laboratory assistants.

5) Not all people who use scientific skills in their work will need to have science qualifications. This is the case if science skills only form a <u>small part</u> of their job (small users — see previous page).

... And a Wide Range of Skills

On top of <u>formal qualifications</u>, everybody who uses science in their work needs other <u>skills</u> — the exact skills required will depend on the <u>nature</u> of the job, but they might include things like:

1) <u>Research skills</u> — finding <u>information</u> from books, scientific journals or the internet.

2) <u>Communication skills</u> — getting your ideas across in a <u>clear</u> way.

3) <u>Numeracy skills</u> — being able to take <u>measurements</u>, carry out <u>calculations</u> and <u>analyse</u> data using <u>statistics</u>.

4) <u>IT skills</u> — using computer packages, e.g. to make <u>spreadsheets</u> and <u>databases</u>.

5) <u>Planning skills</u> — planning <u>investigations</u> that will be <u>successful</u> and hopefully give <u>good results</u>.

6) <u>Analytical skills</u> — breaking problems down into <u>smaller</u>, easier-to-solve chunks.

7) <u>Observational skills</u> — making <u>accurate</u> and <u>useful</u> observations of experiments and accurately <u>recording</u> results.

8) <u>Applying specialist knowledge</u> — <u>assessing</u> results and drawing <u>conclusions</u>.

9) <u>Team working skills</u> — working as a team is important for a lot of scientific work. Good team working skills will mean that the task can be completed to a <u>high standard</u> in an <u>efficient way</u>.

A typical team of scientists at work — remember, there's no 'I' in 'team'.

Next time you break one of your mam's vases...

...just tell her you were practising your analytical skills. You'll probably need skills like this no matter what kind of <u>job</u> you go for. But you're probably wondering just <u>why</u> you need to know about all this — well, one of the things you have to write about in your <u>report</u> is the <u>skills</u> and <u>qualifications</u> of scientists, so it's going to come in mighty <u>handy</u> for that. Also I thought that maybe you'd just like to know.

Report: Science in the Workplace 1

Well, now the section's over it's time to crack on with that <u>report</u> I've been blabbering on about.

You Need to Write a Report on Science in the Workplace

This will be the FIRST OF TWO reports that make up your portfolio for <u>UNIT 1: SCIENCE IN THE WORKPLACE</u>.

Your report will have <u>two bits</u> to it:

1) A <u>DESCRIPTION</u> of a <u>minimum</u> of <u>three</u> organisations that use science or scientific skills (and a more <u>in-depth study</u> of one of them), including:
- <u>General information</u> about the <u>organisation</u> — what <u>products</u> they make (or what <u>services</u> they provide), where they're <u>located</u>, and whether they're <u>local</u>, <u>national</u> or <u>international</u>.
- The <u>jobs</u> of those employed, and what <u>qualifications</u> and <u>skills</u> they have.

2) You also need to write about the <u>TYPES OF CAREERS</u> that are available in science. (You could link this to your workplace report by identifying the careers in the organisations you studied.)

Choose Your Organisations Carefully

1) It's no good picking three organisations that are <u>all the same</u> (e.g. three international drug companies) — it'd be pretty dull for you and won't get you great marks. Try to pick organisations from <u>different areas of science</u> (e.g. healthcare, environment and engineering) and try to pick one <u>local</u>, one <u>national</u> and one <u>international</u> organisation.

2) There are plenty of places you can look for <u>inspiration</u> — e.g. the phone book, the internet, newspapers (local and national), job adverts and the local job centre. Your friends and family might have some good ideas (or might even work for an organisation that uses science).

Find Out General Information from Websites

1) If they're a biggish company they'll probably have a <u>website</u>. This should tell you loads of the things you need to know, e.g. <u>what they do</u> and <u>where they're</u> <u>located</u> (and if it's in more than one country you know they're international).

2) If you can't find information about your chosen organisation on the internet you might have to <u>write</u> and ask for it. You could also prepare a <u>questionnaire</u> and send it to the organisation.

3) You'll get better marks if you <u>describe</u> things, rather than just stating them — don't just say, 'They make drugs' — instead describe what type of drugs they make and what they're used for etc.

For top marks you also need to:
- <u>explain why</u> the organisation is <u>located where it is</u>,
- <u>explain</u> its <u>importance</u> to <u>society</u>,
- <u>describe</u> how it <u>affects</u> the <u>local environment</u>.

Then Find Out About Science Careers

There are plenty of ways to find out about science careers, e.g. <u>job / career websites</u>, <u>company websites</u> (under the job / career link), your <u>local job centre</u>. It's worth looking out for <u>job descriptions</u> — these tell you what qualifications and skills you have to have to do that job. Some big employers and science institutions have dedicated careers websites, e.g. the NHS have www.nhscareers.nhs.uk.

Writing reports? — I thought this was Science not English...

There's loads of info out there — the hardest part is knowing <u>where to start</u>. If you're really struggling to find anything about a particular organisation early on then it might be better to pick a different one.

Avoiding Hazards

Scientific work can be <u>dangerous</u>. You need to be able to work <u>safely</u> in order to <u>prevent</u> accidents from happening. This applies to all workplaces, e.g. school and industrial labs.

There are Six Main Types of Hazard You Should be Aware Of

Hazards need to be <u>identified</u> so that they can be <u>avoided</u>. So, first things first — what types of hazard are there?

Hmm... Where did my bacteria sample go?

1) <u>MICROORGANISMS</u> — these are a particular problem in <u>microbiology labs</u>. The biggest hazard is coming into contact with microorganisms that can <u>cause disease</u>, e.g. <u>viruses</u> and <u>bacteria</u> (see page 29 for more).

2) <u>RADIATION</u> — this is emitted by <u>radioactive materials</u>. The effect on body tissues can be devastating (<u>nausea</u>, <u>weakened immune system</u>, even <u>death</u>) so you need to take precautions (see p.33). This <u>hazard symbol</u> is used to label a radioactive source.

3) <u>CHEMICALS</u> — There are different types of hazardous chemical, each with a <u>hazard warning symbol</u>:

 <u>Oxidising</u> — These provide <u>oxygen</u>, which allows other materials to <u>burn more fiercely</u>, e.g. liquid oxygen.

 <u>Toxic</u> — Can cause <u>death</u> either by being <u>swallowed</u>, <u>breathed</u> in, or <u>absorbed</u> through the skin, e.g. cyanide.

<u>Flammable</u> — <u>Catch fire</u> easily, e.g. petrol.

<u>Corrosive</u> — <u>Attacks and destroys materials</u>, particularly <u>living tissues</u> such as eyes and skin, e.g. sulfuric acid.

<u>Irritant</u> — Not corrosive but can cause <u>reddening or blistering</u> of the skin, e.g. bleach.

4) <u>ELECTRICITY</u> — electrical hazards include long or frayed cables, cables touching something hot or wet, damaged plugs, overloaded sockets and machines without covers. Electrical hazards can cause <u>electric shocks</u> that may lead to <u>burns</u> or <u>death</u>.

5) <u>GAS</u> — it's important to make sure all gas hoses and taps are in <u>proper working order</u>. Gas is <u>flammable</u>, but thanks to its <u>smell</u> it's usually pretty obvious if someone leaves a tap on. However, left unnoticed it can cause <u>suffocation</u> or an <u>explosion</u>.

6) <u>FIRE</u> — many things in the workplace can cause fires, particularly in the petrochemical industry. <u>Damaged electrical appliances</u> are a big culprit. For more on fire see page 10.

Safety Signs Warn You of Hazards in the Workplace

<u>Safety signs</u> give health and safety information in the normal course of work. There are <u>four colours</u> of safety sign — they have specific meanings.

 <u>Blue</u> — <u>mandatory</u> sign. Instruction <u>must</u> be followed.

 <u>Yellow</u> — <u>warning</u> sign. Take <u>care</u>.

 <u>Green</u> — <u>safety</u> information. E.g. fire exit / first aid point.

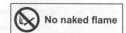 <u>Red</u> — <u>prohibition</u> sign. Action shown must <u>not</u> be carried out.

Risk — not just a thrilling board game...

You'll need this stuff for doing your <u>report</u>, and if you don't follow my advice, then you're just asking for <u>trouble</u>. Don't come crying if you catch <u>hepatitis</u> or <u>electrocute</u> yourself on an overloaded socket.

Avoiding Hazards

Once you've identified your hazards, the next step is to prevent accidents from happening.

The Risk of Injury can be Reduced in Four Ways

The majority of accidents happen because of human error.
The chances of an accident happening are reduced by things like:

1) Proper behaviour — This includes things like not running around the laboratory, and holding scissors or blades in the correct way. It's also important not to eat, drink or smoke in the lab — this is especially important when handling microorganisms and toxic or flammable chemicals.

2) Using equipment properly — all equipment will have instructions and it's important to follow these exactly. Improper use might damage the equipment, but could also lead to serious injury.

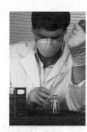

When carrying out any experiment it's important to have the right protective equipment.

3) Using protective and safety equipment — if it's needed, protective equipment has to be provided — it's the law (see below). Lab coats protect your clothes and safety glasses prevent chemicals or flying glass from damaging your eyes. In some scientific workplaces such as hospital laboratories it's important for workers to wear masks and gloves to prevent them being infected with nasty diseases.

4) Following correct procedures — when carrying out an experiment you should always have a well-planned procedure before beginning. It's important to follow the procedures, e.g. using too much of a substance could result in injury if the substance is toxic or flammable.

Workplaces are Governed by Health and Safety Regulations

Health and safety legislation is there to provide employees with a safe and healthy working environment. There's a long list of regulations, which cover things like:

1) General health and safety — the workplace must not be a big risk to the people who work there.

2) Electricity — covering the safe use of electricity (obvious really).

3) Personal protective equipment — employers have to provide protective equipment free of charge where it's needed, e.g. goggles, gloves, helmets etc.

4) Control of hazardous substances — employers are required to label all hazardous substances. They must also have a policy on how the risks of using them can be kept as low as possible.

Protective equipment like helmets and safe footwear must be provided.

> Employers are legally required to assess the risks within the workplace. A risk assessment is an examination of what could cause harm in the workplace. There are five stages to a risk assessment:
> 1) Look for hazards.
> 2) Assess who may be harmed and how.
> 3) Decide what action, if any, needs to be taken to reduce the risk.
> 4) Document the findings.
> 5) Review the risk assessment regularly.

Health and safety officials can enter workplaces at any time to carry out an inspection. Where they find problems they can issue instructions for improvements or stop work being carried out altogether. If breaches of health and safety regulations are really serious employers could end up in court.

It's better to be safe than sorry...

All this might seem pretty dull but you'll be thanking your lucky stars when some rather fetching safety glasses save you from losing your eye in a freak beaker accident.

Radioactive Substances

Many industries use radioactive materials, but they can be <u>dangerous</u> because they <u>produce radiation</u>. This can <u>damage</u> your body (see p.33 for more), so it's pretty important to know how to <u>handle</u> and <u>dispose</u> of these materials in a safe way.

Care Must be Taken When Handling Radioactive Substances

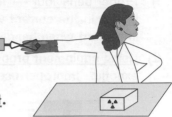

On average, there's one serious incident (resulting in <u>death</u> or <u>serious injury</u>) involving <u>radioactive material</u> in the world each year. The figures are so low because laboratories take <u>precautions</u> when handling radioactive material.

1) <u>Never</u> allow <u>skin contact</u> with a source — always handle with <u>tongs</u>.

2) Keep the source at <u>arm's length</u> to keep it <u>as far</u> from the body <u>as possible</u>.

3) Keep the source <u>pointing away</u> from the body and <u>avoid looking directly at it</u>. (And <u>don't</u> point it at <u>anyone else</u> whilst you're doing this.)

4) <u>Always</u> keep the source in a <u>lead-lined box</u> and put it back in <u>as soon</u> as the experiment is <u>over</u>.

Extra Precautions are Needed for Industrial Workers

1) Radioactive substances are used widely in <u>industry</u> — <u>hospitals</u> use them for things like <u>radiotherapy</u> (p.34) and <u>diagnostic imaging</u>. They're also used for <u>sterilising food</u> and in <u>nuclear power stations</u>.

2) In hospitals workers like <u>radiographers</u> have to keep their <u>radiation exposure</u> to a <u>minimum</u>.
   - They <u>leave the room</u> or stand behind a <u>protective screen</u> when doing things like carrying out radiotherapy (which uses a type of radiation).

STEVE ALLEN / SCIENCE PHOTO LIBRARY

3) Some <u>nuclear power station workers</u> also have to take extra precautions:
   - To prevent workers being exposed to radiation <u>lead-lined suits</u>, <u>lead or concrete barriers</u> and <u>thick lead screens</u> are often used.
   - Some workers have to wear <u>full protective suits</u> to prevent <u>radioactive particles</u> from being <u>inhaled</u> or getting trapped <u>under their fingernails</u> etc.
   - Working in highly radioactive areas is often <u>too dangerous</u> for people so workers use <u>remote-controlled robot arms</u> to move things about. Cool.

Radioactive Material Used in Schools Produces Low-Level Waste

1) The majority of radioactive waste produced by hospitals, universities, schools and colleges is <u>very low-level waste</u>. It usually includes things like gloves, masks, bench coverings, paper towels etc. (Nuclear power stations produce <u>high-level</u> radioactive waste, which has to be dealt with in a special way — see page 59.)

2) Very low-level waste can be sealed in a <u>strong plastic bag</u>, then <u>thrown away</u>. It <u>is</u> safe to dispose of this waste in landfill sites because the level of radiation is so <u>low</u>.

3) Disposal of radioactive <u>sources</u> (the material that's actually used for experiments) is slightly different. Solid sources (like the ones you'll use at school) should be put in a <u>small container</u> and filled up with <u>plaster of Paris</u>. The container <u>should not</u> be labelled to show that it contains a radioactive source (if you labelled it some nosey parker might pick it up). The container can then be put in with normal rubbish (but you can't dispose of sources this way more than once a week).

Robot arms — my preferred dancing style...

It might sound a bit careless just to put radioactive waste out for the bin men but it's only <u>very</u> low-level waste. <u>Nuclear power plants</u> have to be extra <u>careful</u> with their waste. They can't just go around putting weapons grade plutonium in plastic bags and chucking them in the bin — that'd just be wrong.

First Aid

Now, if you've been <u>paying attention</u> over the last few pages then hopefully you'll <u>never need</u> to use what you learn on this page. Having said that, there's always going to be some <u>idiot</u> clowning around, <u>causing trouble</u> for the rest of us — so it's probably for the best if you have a <u>good read</u> over this stuff anyway.

First Aiders Could be the Difference Between Life and Death

It's a good idea, but <u>not a legal requirement</u>, to have as many people as possible trained in basic <u>first aid</u>. They could be vital in saving the life of a person in <u>any situation</u>, e.g. at work, in the street or at home.

Training courses in basic first aid are provided by <u>St John Ambulance</u>, <u>St Andrew's Ambulance Association</u>, and the <u>British Red Cross</u>. All these organisations have websites and can be found in the phone book.

In Any First Aid Situation Follow a Clear Plan of Action

This will stop you placing <u>yourself</u> in danger and will help you to respond in the <u>right way</u>:

1) <u>Assess the situation</u> — what has happened? Is anyone still in danger?
2) <u>Make the area safe</u> — protect yourself and the casualty from danger.
3) <u>Give emergency aid</u> — give appropriate first aid (see below). If there's more than one casualty, the ones with <u>life-threatening</u> conditions should be treated <u>first</u>.
4) <u>Get help</u> — once the casualty has been <u>stabilised</u> call an <u>ambulance</u>.

You Need to Know the Treatment for Common Injuries

There are <u>seven</u> common injuries that you might encounter in the laboratory. You need to know <u>what they are</u>, <u>what basic first aid should be given</u> and when it would be <u>unsafe to give first aid</u>.

1) <u>Heat burns and scalds</u> — if they're <u>minor</u> flood the injured part with <u>cool water</u> for at least 10 minutes, then cover with a <u>sterile dressing</u>. If they're <u>pretty serious</u> cool, damp cloths should be used instead and you should ring an <u>ambulance</u>.
2) <u>Chemical burns</u> — flood the injured part with water for at least 20 minutes, remove any <u>contaminated clothing</u> and arrange for the casualty to be sent to <u>hospital</u>. You should <u>not</u> attempt to give first aid if there are <u>chemical fumes present</u> or if there has been significant <u>chemical spillage</u>.
3) <u>Poisoning due to fume inhalation</u> — the priority is to get the casualty into <u>fresh air</u> and to get <u>medical help</u>. You should not attempt to move the casualty if there are <u>fumes in the area</u>.
4) <u>Poisoning due to swallowing</u> — the casualty needs to go straight to <u>hospital</u>. Never attempt to make the casualty <u>vomit</u> and never give them anything to <u>drink</u> (although <u>small sips</u> of water are OK if they've swallowed something <u>corrosive</u>).
5) <u>Electric shock</u> — turn off the <u>electrical supply</u> before doing anything else (don't touch the victim until you've done this), then ring an <u>ambulance</u>. If the casualty stops breathing you need to be prepared to give <u>rescue breathing</u> (mouth-to-mouth resuscitation) and <u>chest compressions</u>.
6) <u>Cuts</u> — clean the wound under <u>running water</u>, raise the injured part if possible and apply a <u>dressing</u> (<u>don't</u> try to pull objects out of wounds — pad around them and bandage over the top, then send the casualty to <u>hospital</u>).
7) <u>Particles or chemicals in the eye</u> — particles or chemicals should be <u>flushed out</u> of the eye using lots of <u>sterile water</u>. You need to do this for 10 minutes, then bandage the eye before sending the casualty to <u>hospital</u>.

First aid — relief for someone in need of a brew...

An important thing to remember about first aid is that you shouldn't do anything that might put <u>you</u> or the <u>casualty</u> at <u>risk</u>. Never do anything that you're <u>not sure about</u> just for the <u>sake</u> of doing something.

Fire Prevention

Fires are responsible for many deaths every year, but only 6% of these occur in the workplace. The low figure is all thanks to things like <u>fire instructions</u>, <u>sprinklers</u> and <u>extinguishers</u>...

Fire Instructions Tell You What To Do in the Event of a Fire

Fire instruction notices should be displayed at <u>prominent</u> points in a building — they tell you the <u>quickest route</u> to leave the building and <u>where to assemble</u>. Make sure you're <u>familiar</u> with your school lab ones. Also make sure you know what the fire alarm <u>sounds like</u> — it should be tested at least <u>once a week</u>.

If the fire alarm <u>sounds</u> you should:

1) Leave the building by the <u>quickest escape route</u> — <u>never</u> use lifts, escalators or revolving doors.

2) Go to the designated <u>assembly point</u> and wait there until a fire warden takes a roll-call — <u>don't</u> wander off or go home, or somebody might <u>re-enter</u> a burning building to <u>look for you</u>.

If you <u>discover</u> a fire you should:

1) <u>Sound</u> the fire alarm (usually by smashing the glass at a fire alarm point).

2) <u>Call the fire brigade</u> (though some alarms will automatically alert the fire brigade).

3) If the fire is <u>small enough</u>, use a hand-held extinguisher to tackle it (the different types of extinguisher are listed below) — but <u>never</u> put yourself at <u>risk</u> in attempting to put a fire out, and <u>always</u> stand <u>between</u> the fire and your escape route.

4) Leave the building by the <u>quickest escape route</u> and report to the <u>assembly area</u>.

Fire Doors and Sprinkler Systems can Stop Fire Spreading

Fires can spread easily through <u>open areas</u> such as <u>corridors</u> and <u>stairwells</u>. There are <u>two</u> common features installed in the workplace to <u>slow down</u> the spread of fire.

1) <u>Fire doors</u> act as barriers to hold back <u>smoke</u> and <u>flames</u>. Fire doors must be kept <u>shut</u> at all times or be fitted with <u>automatic closing devices</u> — <u>never</u> wedge a fire door open.

2) <u>Sprinkler systems</u> are usually installed in <u>high-risk</u> areas, such as <u>storerooms</u>. They're very effective at <u>containing fires</u> — they spray water from the ceilings. But, they're <u>expensive</u> to install and maintain, they need a water supply at <u>high pressure</u>, and if there's a minor fire (which can be put out with a fire extinguisher) they can cause a lot of unnecessary <u>mess</u> and <u>damage</u> to equipment and stock.

Know Which Type of Fire Extinguisher to Use

There are <u>six types</u> of hand-held fire extinguisher. All new fire extinguishers are painted red with a <u>colour-coded</u> band or panel to identify its <u>contents</u> and the <u>type of fire</u> it can be used on. It's important to use the <u>right fire extinguisher</u> for the type of fire — you could make the <u>fire worse</u> if you use the <u>wrong one</u>.

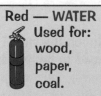

Red — WATER
Used for: wood, paper, coal.

Black — CARBON DIOXIDE
Used for: wood, paper, coal, liquids, electrical equipment.

Green — VAPORISING LIQUID
Used for: liquids, electrical equipment.

You can also use a <u>fire blanket</u> to smother a fire if it's small and self-contained.

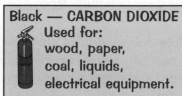

Cream — FOAM
Used for: wood, paper, coal, liquids.

Blue — DRY POWDER
Used for: wood, paper, coal, liquids, gases, electrical equipment.

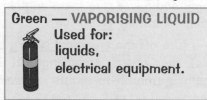

Yellow — WET CHEMICAL
Used for: cooking oil and fats.

Fire doors — bet they have hot handles...

Remember — it's dead important to use the <u>right kind</u> of extinguisher for a fire. You should <u>never</u> use <u>water</u> on an <u>oil fire</u> (e.g. a chip pan fire) — it could make the fire even <u>worse</u> and cause an <u>explosion</u>.

Report: Science in the Workplace 2

Now on to the unpleasant subject of <u>assessment</u> — those lovely examiners want you to write a report on working safely in the workplace. It's probably a good idea to base this report on the <u>same</u> organisation you picked for your first report.

You Need to Write a Report on Working Safely

This will be the SECOND OF TWO reports that make up your portfolio for <u>UNIT 1: SCIENCE IN THE WORKPLACE</u>.

Your report will have <u>three bits</u> to it:

1) A description of how <u>HAZARD ASSESSMENT</u> is managed for a <u>scientific workplace</u>.
2) Details about <u>FIRST AID</u> for that workplace.
3) Details about <u>FIRE PREVENTION</u> for that workplace.

Find Out How They Assess Hazards

1) To get <u>specific information</u> about health and safety in your chosen scientific workplace you'll need to write a short <u>questionnaire</u>, <u>phone them</u> or <u>visit in person</u>.

2) By law, firms with five or more employees must have a written <u>health and safety policy</u> — you could try asking (politely, of course) for a copy. It should contain a lot of the information you're after.

3) There are plenty of <u>questions</u> you could ask to get the information you need:

- <u>Risk assessment</u> — who's responsible for making risk assessments? <u>How often</u> are risk assessments reviewed? (You could include a copy of a risk assessment in your report.)
- <u>Machinery and equipment safety</u> — what's the procedure for checking electrical and mechanical equipment? Is there a maintenance logbook?
- <u>Hazardous substances</u> — what substances are used? <u>Who's exposed</u> to these substances? What <u>storage</u> arrangements and <u>safety precautions</u> are in place?
- <u>Protective clothing and equipment</u> — what items of protective clothing are provided? When should these items be used? (You should find this information in the risk assessments.)
- <u>Safety information</u> — what information on health and safety is available to staff?
- <u>Job training</u> — is there a safety induction programme?
- <u>Safe working practices</u> — How do employers make sure health and safety procedures are followed?

Then Find Out About First Aid and Fire Safety

There are plenty of questions you could ask to get this information:

- <u>First aid</u> — where are the <u>first aid boxes</u>? Where is the <u>accident book</u> kept? What is the policy on having trained <u>first-aiders</u>?
- <u>Fire safety</u> — what are the <u>emergency evacuation procedures</u>? What <u>hand-held</u> fire-fighting equipment is available?

For top marks you also need to:
- <u>research</u> working safely in your <u>school or college</u> (do all the things above again),
- <u>compare</u> this to your chosen <u>workplace</u> (point out how they are <u>different</u> and how they are <u>similar</u>).

Reports, investigations? Who do you think you are? — Poirot?

There's no magic number of words that your report must be — it has to be long enough to get your <u>point across clearly</u>, but not so long that you fill it with <u>waffle</u>. Good luck.

Organisms and Cells

All living things are made from <u>cells</u> — they're the very basic <u>'building blocks'</u> of life. You, me and that dying plant in the corner of your bedroom are all made up from cells. Getting your head around what cells are and how they work is the first step in understanding nearly everything that goes on in your body. Everything from breathing to regulating body temperature requires the cooperation of your cells.

Animal Cells have Three Main Parts

Nearly all cells contain the <u>same features</u> which are important in keeping the cell alive and allowing it to function properly.

A human being is made up of <u>trillions</u> of cells...

...and each cell has <u>three</u> important features you need to learn about:

1) NUCLEUS

Contains <u>genetic material</u> that controls the activities of the cell (see page 24).

2) CYTOPLASM

A gel-like substance where most of the <u>chemical reactions</u> happen.

3) CELL MEMBRANE

Holds the cell together and controls what goes <u>in</u> and <u>out</u> (see page 14).

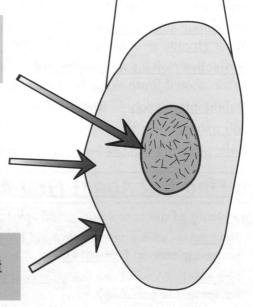

Cell — a stockbroker's favourite word...

Cells don't usually work alone — they team up to make tissues, organs and organ systems.
So, a group of <u>similar cells</u> is called a <u>tissue</u>. A group of <u>different tissues</u> form an <u>organ</u> and a <u>group of organs</u> working together form an <u>organ system</u>. For example, loads of muscle cells make up muscle tissue... which with other tissues makes up the heart (an <u>organ</u>). The heart and other organs (like blood vessels) together make up the circulatory system (an organ system — see page 17 for more).
There's one extra stage to this — lots of organ systems make up an <u>organism</u>... fascinating.

Specialised Cells

Most cells are specialised for a specific job, and in the exam you might have to explain why the cell they've shown you is so darn good at what it does. Well, it's a lot easier if you've already learnt them...

Red Blood Cells Carry Oxygen Around the Body

The structure of a red blood cell is adapted to its function — carrying oxygen around the body:

1) Red blood cells are small and have a biconcave shape (which is a posh way of saying they look a little bit like doughnuts — see the diagram) to give a large surface area for absorbing and releasing oxygen.

2) They contain a substance called haemoglobin, which absorbs oxygen.

3) Red blood cells don't have a nucleus — this frees up space for more haemoglobin, so they can carry more oxygen.

4) Red blood cells are very flexible. This means they can easily pass through tiny blood vessels.

There's loads more about how red blood cells work on page 16. It's quite exciting really.

(Actually, maybe I need to get out more.)

White Blood Cells Help Fight Disease

White blood cells are also adapted to their function — their main role is defence against disease.

1) They have a flexible shape, which helps them to engulf any microorganisms they come across inside the body. Basically the white blood cell totally surrounds the microorganism and then digests it.

2) They can also produce antibodies to fight microbes, and antitoxins to neutralise the microbes' toxins (poisons).

Nerve Cells Carry Messages Around the Body...

Neurones (nerve cells) transmit information as electrical impulses around the body. The electrical impulses pass along the axon of the cell. These cells are specialised in several ways...

Here's a typical nerve cell:

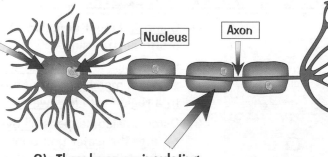

See pages 19-20 for more on nerves.

Cell body | Nucleus | Axon

1) They're long, which speeds up the impulse as it doesn't have to keep 'jumping' between neurones.

2) They have an insulating sheath. This acts as an electrical insulator, which also speeds up the impulse.

3) They have branched endings. This allows them to connect to lots of other neurones.

Blood lolly — special-iced cells...

There are loads more specialised cells in the body apart from these three, e.g. kidney, liver and muscle cells. Sketch out the diagrams for the ones on this page and point out their special features till you know them.

Diffusion and Osmosis in Cells

Particles need to <u>move in and out</u> of cells so that your body can carry out reactions like respiration (see next page). This is basically how it happens...

Cell Membranes Control Which Substances Can Get In and Out

1) Cell membranes are kind of clever because they hold everything <u>inside</u> the cell, but they let stuff <u>in and out</u> as well.

2) Only very <u>small particles</u> can get through cell membranes though (e.g. things like <u>glucose</u>). This happens by <u>DIFFUSION</u>, and it's really simple — it's just the <u>movement</u> of particles from places where there are <u>lots</u> of them to places where there are <u>fewer</u> of them.

3) Particles actually pass <u>both ways</u> through the membrane.

4) But because there are <u>more</u> particles on one side than the other there's a steady <u>net flow</u> into the region with <u>fewer</u> particles.

5) This causes the region with less particles to <u>fill up</u> with more. So, diffusion tends to "<u>even up</u>" the amount of particles on either side of the membrane.

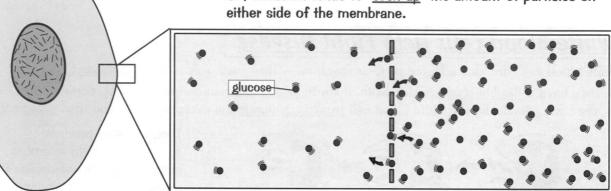

glucose

6) <u>Diffusion</u> happens <u>all over the body</u>:
   - E.g. the <u>food</u> we eat is digested — it <u>diffuses</u> from the <u>digestive system</u> into the <u>blood</u> and then diffuses out of the blood at the <u>cells</u>.
   - E.g. the <u>oxygen</u> in the air we breathe in <u>diffuses</u> from the <u>lungs</u> into the <u>blood</u>.

Osmosis is Diffusion of Water Molecules

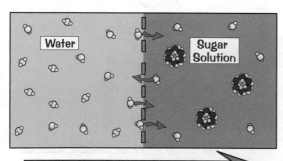

Water

Sugar Solution

Net movement of water molecules

1) <u>Osmosis</u> is just diffusion of water molecules... pretty easy really.

2) So <u>osmosis</u> is the <u>movement</u> of <u>water molecules</u> across a <u>membrane</u> from a region with <u>more water</u> to a region with <u>less water</u>.

3) Osmosis also happens <u>all over</u> the <u>body</u>:
   - E.g. the <u>water</u> you drink moves from the <u>digestive system</u> into the <u>blood</u> by <u>osmosis</u>.
   - E.g. water moves from the <u>blood</u> into the <u>cells</u> by osmosis.

Osmosis is a special case of diffusion, that's all...

So, if <u>osmosis</u> is the <u>same</u> as <u>diffusion</u> but with <u>water</u>, why don't they just call it water diffusion. It's not like it's even a good word — how can you possibly <u>make a joke</u> using osmosis. Not even my panel of specially trained, professional joke thinker-uppers can come up with anything remotely amusing about it.

Respiration

Respiration is a <u>really important process</u> — it <u>releases</u> the <u>energy</u> we need from all that <u>food</u> we eat. Without respiration you wouldn't be able to live, let alone revise. So learn this page and thank respiration for being here.

Respiration is NOT "breathing in and out"

See page 18 for breathing in and out.

1) Respiration is NOT breathing in and breathing out, as you might think.
2) <u>Respiration</u> actually goes on in <u>every cell</u> in your body.
3) It's the process of releasing <u>energy</u> from <u>glucose</u> (i.e. from your food).
4) This energy is then used to do things like:

Build up <u>larger</u> <u>molecules</u> (like proteins),

Contract <u>muscles</u>,

Maintain a steady <u>body</u> <u>temperature</u>.

5) All living things "<u>respire</u>", e.g. plants <u>respire</u> too. The only difference between us and them is that they make their food (using light) and we eat ours.
6) Here's a nice <u>definition</u> to learn...

> **RESPIRATION is the process of RELEASING ENERGY from GLUCOSE, which goes on IN EVERY CELL.**

7) You also need to learn the <u>equation</u>:

> **Glucose + Oxygen → Carbon Dioxide + Water (+ Energy)**

So, For Respiration Your Cells Need Glucose...

1 The sugary <u>food</u> you eat is <u>digested</u> into <u>glucose</u> and <u>diffuses</u> from the <u>digestive system</u> into the <u>blood</u>.

2 The <u>blood</u> then <u>carries</u> <u>glucose</u> to the <u>cells</u>.

... And Oxygen

1 Breathing in moves <u>oxygen</u> into your <u>lungs</u> (see page 18).

2 <u>Oxygen</u> then <u>diffuses</u> into your <u>blood</u> (see page 14).

3 <u>Blood</u> moves around the <u>circulatory system</u> (see page 17) to all <u>cells</u>, <u>supplying</u> the oxygen they need for <u>respiration</u>.

Respiration — it's a busy life...

Isn't it strange to think that each individual living cell in your body is respiring every second of every day — releasing energy from the food you eat. Next time someone accuses you of being lazy you could claim that you're busy respiring — it's enough to make anyone feel tired.

The Blood

Blood — it's very useful stuff. Its main use is to make you look cool when you fall off your bike, but it's also pretty good at <u>carrying things around the body</u>...

Blood is Made Up of Four Main Things

The blood is basically a big <u>transport system</u> for moving <u>substances</u> to and from body cells.
It's also involved in fighting off microorganisms and stopping you bleeding to death when you cut yourself.
You need to learn the four main bits that help it do this:

1) Red Blood Cells Carry Oxygen

The job of the red blood cells is to transport <u>oxygen</u> from the <u>lungs</u> to all the <u>cells in the body</u> — and as you saw on page 13 they're well <u>adapted</u> to do just that.

1) Remember — red blood cells contain <u>haemoglobin</u>.
2) In the lungs haemoglobin reacts with oxygen to form <u>oxyhaemoglobin</u>.
3) In the body cells the <u>reverse</u> happens and oxyhaemoglobin <u>releases oxygen</u>.

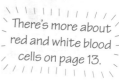

There's more about red and white blood cells on page 13.

2) White Blood Cells Fight Disease

1) White blood cells play a really important role in <u>protecting your body</u> against attack from <u>microorganisms</u>.
2) They travel around the blood and crawl into every part of you, constantly <u>patrolling</u> for microorganisms. When they come across an invading microorganism they <u>engulf</u> it (see page 31).

3) Platelets Help Blood Clot

1) These are <u>small fragments</u> of <u>cells</u> that have <u>no nucleus</u>.
2) They help the blood to <u>clot</u> at the site of a wound. This stops all your <u>blood pouring out</u> and stops <u>microorganisms</u> getting in. (So basically they just float about waiting for accidents to happen!)

4) Plasma is the Liquid Bit

Plasma is a pale yellow liquid which carries <u>just about everything</u>:

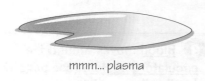

mmm... plasma

1) <u>Red blood cells</u>, <u>white blood cells</u> and <u>platelets</u>.
2) <u>Water</u>.
3) Digested food products like <u>glucose</u>. These are absorbed from the gut and taken to all body cells.
4) <u>Carbon dioxide</u> from the body cells to the lungs.
5) <u>Hormones</u> — these act like chemical messengers (see page 22).
6) <u>Antibodies</u> and <u>antitoxins</u> produced by the white blood cells (see pages 13 and 31).

Platelets — good for small dinners...

And you thought blood was just red and wet — turns out there's a lot more to it. It's important not just to <u>learn</u> what the <u>different parts</u> of blood are — also make sure you know <u>what</u> they all do.

The Circulatory System

Blood doesn't just move around the body on its own, of course. It needs a pump and tubes to flow through, which is where the heart and the blood vessels come in. The blood, blood vessels and heart make up the circulatory system — its main function is to get food and oxygen to every cell in the body.

Blood is Carried Around the Body by Vessels

There are three different types of blood vessel:

1) ARTERIES — these carry the blood away from the heart.

2) CAPILLARIES — these are small vessels involved in the exchange of materials at the tissues.

3) VEINS — these carry the blood to the heart.

Normally, arteries carry oxygenated blood (blood with oxygen) and veins carry deoxygenated blood (blood without oxygen).

The pulmonary artery and pulmonary vein are the big exceptions to this rule (see diagram below).

The Blood is Pumped Around the Body by the Heart

1) The heart is a pump — it supplies the force needed to push the blood all round the body through the blood vessels. To every last tissue and back.

2) Humans have a double circulatory system. This means that there are two circuits of blood vessels — one going to the lungs and one to the rest of the body:

1) The first circuit connects the heart to the lungs. Blood is pumped to the lungs to take in oxygen. The blood then returns to the heart.

2) The second circuit pumps the oxygenated blood around the body through the arteries.

3) The arteries eventually split off into thousands of tiny capillaries which take blood to every cell in the body.

4) The blood gives up its oxygen to the cells.

5) The veins then collect the deoxygenated blood and carry it back to the heart to be pumped out to the lungs again.

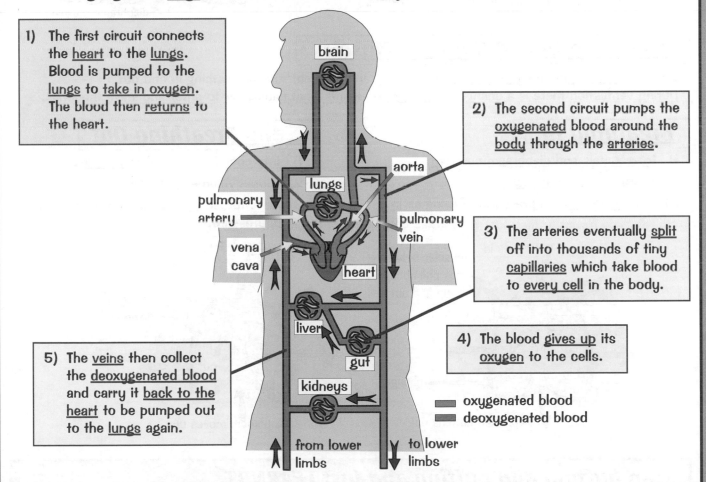

oxygenated blood
deoxygenated blood

Okay — let's get to the heart of the matter...

The human heart beats 100 000 times a day on average. You can feel a pulse in your wrist or neck (where the blood vessels are close to the surface). This is the blood being pushed along by a heart beat. Don't forget that the circulatory system also delivers glucose to the cells, for respiration.

Breathing

You need to get air (containing <u>oxygen</u>) into your lungs so the oxygen can diffuse into the blood... which is where <u>breathing</u> comes in. (They sometimes call it '<u>ventilation</u>' in the exams — don't get confused with those big shiny metal things that Bruce Willis likes climbing through, it's just breathing, OK.)

The Thorax — The Top Part of Your Body

The <u>thorax</u> is the part of the 'body' from the neck down to the diaphragm. There are a few parts you need to know...

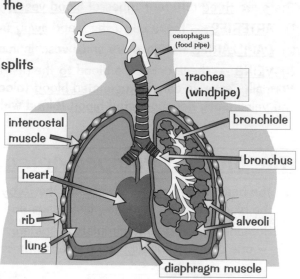

1) The <u>lungs</u> are like big pink <u>sponges</u>.

2) The <u>trachea</u> (the tube from your mouth to your lungs) splits into two tubes called '<u>bronchi</u>', one goes to each lung.

3) The bronchi split into progressively smaller tubes called <u>bronchioles</u> that end with small bags called <u>alveoli</u> — this is where gas exchange takes place.

4) The <u>ribs</u> protect the lungs and the heart etc. They're also important in breathing (see below).

5) The <u>intercostal muscles</u> are the muscles in between the ribs.

6) The <u>diaphragm</u> is the large muscle at the bottom of the thorax — it's also important for breathing.

Breathing In and Out Uses Muscles

Both the <u>diaphragm</u> and <u>intercostal muscles</u> play an important role in breathing in and out. During ventilation there is a change in <u>pressure</u> — this is what causes air to enter and leave the lungs.

Breathing In...

1) <u>Intercostals</u> and <u>diaphragm contract</u>.
2) Thorax volume <u>increases</u>.
3) This <u>decreases</u> the pressure, drawing air <u>in</u>.

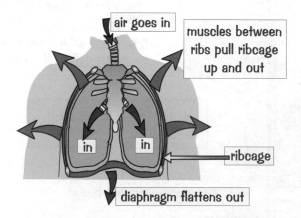

...and Breathing Out

1) <u>Intercostals</u> and <u>diaphragm relax</u>.
2) Thorax volume <u>decreases</u>.
3) Air is forced <u>out</u> because of an <u>increase</u> in pressure.

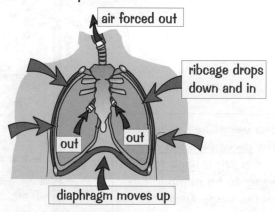

Stop huffing and puffing and just LEARN IT...

If you've ever fancied a career in the <u>medical profession</u> then you'll need to know this stuff inside out. Plus it comes in really handy in everyday life — I regularly drop interesting biology facts into conversation in an attempt to woo the opposite sex. Even if it doesn't go down too well at least it's stuck in your head that air is drawn <u>into the lungs</u> because of a <u>decrease in pressure</u> caused by an <u>increased thorax volume</u>. Oh, and it'll also be useful in the exam. So many reasons to learn this page.

The Nervous System

The job of the nervous system is to <u>detect stimuli</u> — <u>changes</u> in your <u>environment</u> (i.e. what's going on <u>around you</u>). It then <u>decides</u> what to do about the stimuli and <u>makes it happen</u>, e.g. if you see a ball coming at your head your nervous system detects it and makes you duck... pretty useful really.

Sense Organs Contain Receptors That Detect Stimuli

There are five different <u>sense organs</u>, which all contain different <u>receptors</u>.

Receptors are groups of cells which are sensitive to a <u>stimulus</u> such as light or heat, etc.

When they're activated by a stimulus they generate <u>electrical impulses</u>.

<u>Sense organs</u> and <u>Receptors</u>
Don't get them mixed up:

The <u>eye</u> is a <u>sense organ</u> — it contains <u>light receptors</u>.

The <u>ear</u> is a <u>sense organ</u> — it contains <u>sound-receptors</u>.

The <u>Five Sense Organs</u> and the <u>stimuli</u> that each one is <u>sensitive to</u>:

1) Eyes
<u>Light</u> receptors.

2) Ears
<u>Sound</u> and '<u>balance</u>' receptors.

3) Nose
<u>Taste</u> and <u>smell</u> receptors (chemical stimuli).

4) Tongue
<u>Taste</u> receptors:
Bitter, salt, sweet and sour (chemical stimuli).

5) Skin
<u>Touch</u> and <u>temperature</u> receptors.

There's more about how these all work together on the next page.

Other Parts of the Nervous System

Spinal cord
This is a bundle of nerves that transmits information up and down the body.

Motor Neurones
The <u>nerve cells</u> that carry signals from the brain to the <u>effectors</u> (muscles and glands).

Sensory neurones
The <u>nerve cells</u> that carry signals from the <u>receptors</u> in the sense organs towards the brain.

Effectors
All your <u>muscles</u> and <u>glands</u> will respond to nervous impulses.

Hurrah for sense organs — without them you couldn't revise...

Well, maybe not hurrah then. In a <u>single cubic centimetre</u> of your <u>brain</u> you can have well <u>over 50 million</u> nerve cells. Each one of these exciting little cells can communicate with <u>thousands</u> of other nerve cells to process information from the <u>receptors</u> — and you thought computers were impressive.

The Nervous System

You have <u>millions</u> of neurones (nerve cells) in your body that go all over the place. <u>Information</u> from receptors is transmitted around the body by neurones as <u>electrical impulses</u>. This is how information about what's going on in the world gets all the way to your big brain.

Neurones Transmit Information as Electrical Impulses

Electrical impulses need to pass along neurones to get to and from the brain...

1) Electrical impulses are <u>generated</u> at one end by <u>receptors</u>.

2) The impulses then travel along the <u>axon</u> of the cell to the other end.

3) They can then be transmitted to other <u>connecting</u> neurones — over a 'gap' called the <u>synapse</u>...

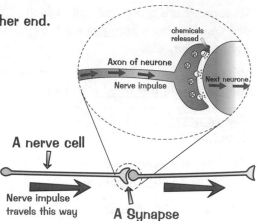

Synapses Connect Neurones

1) The <u>connection</u> between <u>two neurones</u> is called a <u>synapse</u>.

2) The electrical signal is <u>transferred</u> to the next neurone by <u>chemicals</u> which <u>diffuse</u> across the gap.

3) These chemicals then set off a <u>new electrical signal</u> in the next neurone.

The Brain Decides What To Do About a Stimulus

The brain's a bit like a <u>giant computer</u> — it <u>processes information</u> and decides what to do about it. Here's what happens when you detect a stimulus: Here's an example:

1) STIMULUS DETECTED by <u>receptors</u> in a <u>sense organ</u>. <u>Electrical impulses</u> are generated.

1) STIMULUS DETECTED The <u>light receptors</u> in your eye spot a <u>big hairy spider</u> getting nearer... eek. <u>Electrical impulse</u> are generated.

Impulses travel along a <u>sensory neurone</u> to the...

2) BRAIN The information is <u>processed</u> and the <u>brain</u> decides <u>what to do</u> about it. It transmits the <u>decision</u> as electrical impulses.

2) BRAIN The information is <u>processed</u> and your <u>brain</u> decides it'd be a <u>good idea</u> to <u>run away</u>. Electrical impulses are sent to your <u>leg muscles</u>.

Impulses travel along a <u>motor neurone</u> to the...

3) EFFECTOR In <u>response</u> to the decision either some <u>muscles</u> contract or some <u>glands</u> secrete something, e.g. sweat or hormones.

3) EFFECTOR In <u>response</u> to the decision the <u>muscles</u> in the <u>legs</u> <u>contract</u>... so you <u>run</u> away.

Don't let this page get on your nerves... (Man, I am soooo funny)

The <u>spinal cord</u> is a <u>vital</u> part of your nervous system — it <u>carries nervous impulses</u> from all parts of your body to your <u>brain</u> and back again. People who suffer a severe spinal cord injury are often <u>paralysed</u> — they <u>can't feel</u> or <u>move</u> limbs because the nerves can't <u>communicate</u> with the brain in the normal way.

Maintaining Body Temperature

Your body also constantly monitors your internal environment (your insides) to make sure they're working properly and you have the right body temperature...

Body Temperature is Around 37 °C

1) The reactions in your body (e.g. respiration — see p.15) work best at about 37 °C.
2) This means that you need to keep your body temperature around this value — within 1 or 2 °C of it.
3) A part of your brain acts as your own personal thermostat. So it's your nervous system that controls your body temperature (see pages 19-20 for more about the nervous system).

The Skin has Two Tricks for Altering Body Temperature

1) When the brain senses changes in the body temperature it sends nervous impulses to the skin.
2) The skin then has two tricks for controlling body temperature:

When You're TOO HOT:

1) More sweat is produced from sweat glands in the skin. The water in the sweat evaporates, taking heat with it. This helps cool you down.

You might also find somewhere cool, e.g. in the shade.

2) Blood vessels (see p.17) close to the skin's surface get bigger in diameter. This means that more blood gets to the surface of the skin. The warm blood then loses some of its heat to the surroundings.

(This is why you look red when you're hot — it's the increased blood flow to the surface of the skin.)

When You're TOO COLD:

You also shiver to try and warm yourself up.

1) Less sweat is produced, so heat isn't lost when the water in it evaporates (because there's not much evaporation going on).

2) Blood vessels close to the skin's surface get smaller in diameter. This means that less blood gets to the surface of the skin. This stops the blood losing its heat to the surroundings.

(This is why you look paler when you're really cold — there's very little blood going to the surface of the skin.)

Sweaty and red — I'm so attractive in the heat...

If you get way too hot you could get heat exhaustion — you feel really tired and a bit sick, and if it's untreated you could die... scary. It's a similar story if you get too cold (the fancy name for this is hypothermia) — you can slip into a coma and die. Getting too cold also isn't great for your fingers, toes and nose — if the blood supply is cut off for too long the cells in the tissues die. This causes frostbite (where the fingers and toes go all black and manky) and it's quite common in mountaineers.

Hormones and Blood Sugar

Hormones aren't just pesky chemicals that make you an awkward teenager — whatever your mother might say. They're actually <u>very useful little things</u> that help your body <u>run like clockwork</u> every single day. They're just another way to <u>communicate information</u> around the body.

Hormones are Chemical Messengers Sent in the Blood

1) Hormones are <u>chemicals</u> produced in various <u>glands</u>.
2) They're released directly into the <u>blood</u>.
3) They're carried in the blood to other parts of the body — so travel at '<u>the speed of blood</u>'.
4) They travel all over the body but <u>only</u> affect <u>particular cells</u> (called target cells) in particular places.
5) They have <u>long-lasting effects</u>.
6) They control things that need <u>constant adjustment</u>.

Hormones and Nerves Carry Messages but Act Differently

Nerves:
1) Use an <u>electrical</u> signal.
2) Very <u>fast</u> message.
3) Act for a very <u>short time</u>.
4) Act on a very <u>precise area</u>.
5) <u>Immediate</u> reaction.

Hormones:
1) Use a <u>chemical</u> signal.
2) <u>Slower</u> message.
3) Act for a <u>long time</u>.
4) Act in a more <u>general</u> way.
5) <u>Longer-term</u> reaction.

Insulin is a Hormone that Controls Blood Sugar Levels

<u>Insulin</u> is produced in the <u>pancreas</u> and controls how much <u>sugar</u> there is in your <u>blood</u>.

1) Eating <u>carbohydrate</u> foods puts a lot of <u>glucose</u> into the blood from the <u>gut</u>.
2) The normal working of cells <u>removes</u> glucose from the blood.
3) Vigorous <u>exercise</u> removes loads of glucose from the blood.
4) Obviously, to keep the <u>level</u> of blood glucose <u>controlled</u> there has to be a way to <u>add or remove</u> glucose from the blood. And this is it:

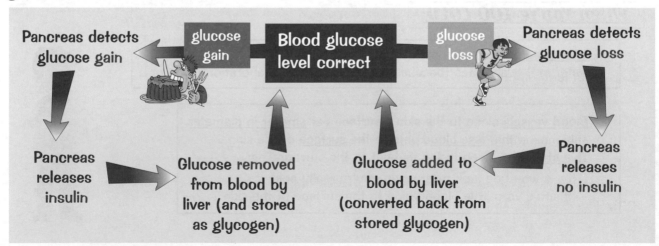

Remember, the <u>addition</u> of insulin <u>reduces</u> blood sugar level.

My blood sugar feels low after that — pass the biscuits...

Some people have a pancreas that <u>doesn't produce any or enough insulin</u>, which can cause their <u>blood sugar to rise</u> to a level that could be <u>fatal</u>. This is a type of <u>diabetes</u>. They often have to <u>inject insulin</u> before meals to reduce their blood sugar levels.

Revision Summary for Section 2.1

That wasn't such a bad section to start Biology on. Now you know how the healthy human body works...
you could almost be a doctor (well, maybe after your A-levels and then five years at Uni, but let's not
dwell on details). If you don't know how it works then you'd better have a look back over these 11 pages
till it's engraved in your brain. Here's a few questions to check that you do know it...

1)   Sketch a cell and label: a) the nucleus, b) the cytoplasm, c) the cell membrane.
2)   What does the cell nucleus contain?
3)   What happens in the cytoplasm?
4)   What does the cell membrane do?
5)   Give three ways in which red blood cells are adapted to their job.
6)   Give two ways in which white blood cells are adapted to their job.
7)   Give three ways in which nerve cells are adapted to their job.
8)   What is diffusion?
9)   Name two places in the body where diffusion takes place.
10) What is osmosis?
11) Name two places in the body where osmosis takes place.
12) Define respiration.
13) What is the energy produced by respiration used for?
14) Give the word equation for respiration.
15) What two substances do body cells need to carry out respiration?
16) Where do each of these substances come from? How do they get to the body cells?
17) What is the compound that absorbs oxygen in red blood cells called?
18) What other type of blood cell is found in the blood? What do they do?
19) What are platelets? What do they do?
20) What is plasma? Name five things that it carries.
21) Name the three different types of blood vessel and describe what each one does.
22) Do arteries normally carry oxygenated blood or deoxygenated blood?
23) What organ pumps the blood around the body?
24) Humans have a double circulatory system. What does this mean?
25) Describe the movement of blood around the circulatory system.
26) Where in the body does the blood pick up oxygen?
27) What is the thorax?
28) Sketch a diagram of everything in the thorax and label the different parts.
29) Describe what happens when you breathe in.
30) Describe what happens when you breathe out.
31) What is a stimulus?
32) Name the five sense organs and the receptors they contain.
33) How do neurones transmit information around the body?
34) Describe how a stimulus brings about a response.
35) What role does the brain play in nervous responses?
36) How does the body reduce its temperature when it's too hot?
37) How does the body increase its temperature when it's too cold?
38) What are hormones? Where are they produced?
39) Explain how insulin controls blood sugar.

Genes and Chromosomes

If you've ever wondered why animals or plants of the same species look or behave slightly differently from each other — you know, a bit taller or a bit fatter — well, it's partly due to their genes. All animals (including humans) are different from each other because their genes are slightly different. Scientists have recently found every single human gene and are now trying to understand what each one does. "But what is a gene?" I hear you cry. Well you're about to find out...

You Need to Know What Chromosomes and Genes Are

Whether you're talking about animals or plants, this basic stuff about genes is pretty much the same...

1) Nearly all cells contain a nucleus. The nucleus contains a plant or animal's genetic information — this is what controls everything a cell does.

2) Genetic information is carried as chromosomes. Chromosomes usually come in pairs. For example, the human cell nucleus contains 23 pairs of chromosomes.

nucleus

A single chromosome

A pair of chromosomes

3) Short sections of a chromosome are called genes. Genes control the characteristics of the body.

4) Each gene contains the instructions needed to make a protein (we say that a gene "codes for a protein"). Proteins determine many characteristics in the body — such as eye colour or whether you are a boy or a girl.

5) In your cells there'll usually be two of each gene (because there are two of each chromosome and each chromosome carries one gene). The two genes may be slightly different from each other — these versions are called alleles of a gene (see page 26).

Genes — they always come in pairs...

This picture's true for most animals and plants (but not for everything — bananas have 11 chromosomes, for example, and things like bacteria carry genetic information slightly differently). Genes are important, because genes control everything a cell does, as well as what characteristics parents pass on to their kids. It's all to do with proteins — genes control the proteins that are made, and proteins control the cell. Got that? Good.

Variation in Plants and Animals

Everyone has <u>different genes</u> from everyone else — except identical twins. Even identical twins can look different though, because it's <u>not just</u> your genes that determine what you look and behave like — the <u>environment</u> (your upbringing) affects you too.

Genes are Responsible for Certain Characteristics

Things like your <u>eye colour</u>, <u>nose shape</u> and <u>blood type</u> are all determined by your <u>genes</u>. There are <u>variations</u> in the genes that cause these characteristics, giving <u>different versions</u> of the <u>same characteristic</u>, e.g. blue eyes or brown eyes. There are <u>two</u> things that cause <u>genetic variation</u>:

1) Sexual Reproduction

There are two stages to sexual reproduction — both can cause genetic variation.

GAMETE FORMATION — MAKING SPERM CELLS AND EGG CELLS
1) Gametes are <u>sperm cells</u> and <u>egg cells</u>.
2) They are formed in the ovaries or testes from <u>reproductive cells</u>.
3) Reproductive cells (like all human body cells) have <u>23 pairs</u> of chromosomes. In each pair there's one chromosome that was <u>originally inherited</u> from <u>mum</u>, and one that was inherited from <u>dad</u>.
4) To make gametes, reproductive cells must <u>split</u> into two and when they do this the genes get <u>shuffled up</u>. Some of your dad's genes are grouped with some from your mum.
5) This shuffling up of genetic material leads to <u>variation</u> in the new generation.

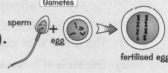

FERTILISATION — JOINING TOGETHER THE GAMETES
1) Fertilisation is when the <u>sperm</u> and the <u>egg</u>, with <u>23 chromosomes each</u>, join to form a new cell with the full <u>46 chromosomes</u> (23 pairs).
2) You never know which two gametes are going to join together, giving <u>even more variation</u>.

2) Mutations

Occasionally a gene may <u>mutate</u> (change) and produce a <u>new characteristic</u>, <u>increasing variation</u>.

Your Environment is Responsible for Other Characteristics

1) <u>Scars</u> — falling off your bike when you're little can leave its mark.
2) <u>Spoken language</u> — which language you learn to speak will depend on how you're brought up.

Most Characteristics are Due to a Mixture of the Two

For some characteristics, it's a <u>combination</u> of genes and environment that causes differences between people...
1) <u>Intelligence</u> — one theory is that your <u>maximum IQ</u> is determined by your <u>genes</u>, but whether you get to it depends on your environment, e.g. your <u>upbringing</u> and <u>school</u> life.
2) <u>Body mass</u> — your <u>natural weight</u> is determined by your genes but can be changed if you <u>diet</u>, or <u>eat loads of junk food</u> and don't do enough exercise.
3) <u>Height</u> — your <u>natural height</u> is determined by your genes, but whether you get to it depends on your environment, e.g. if your growth is stunted in youth by <u>poor diet</u> or if your <u>mum smoked</u> when she was pregnant with you, then you won't be as tall as you could have been.

Well, I've said it all along — sex cells...

So in <u>sexual reproduction</u> a mixture of chromosomes is randomly shuffled into <u>gametes</u>.
Then a random gamete fuses with another random gamete at <u>fertilisation</u> (oh, the romance of it all).

Inherited Characteristics

You inherit things like lovely blue eyes, brown hair and a big nose from your parents
— the problem is that genetic disorders can also be inherited.

Different Versions of Genes are Called Alleles

1) There are <u>two</u> of each <u>chromosome</u> in most of your body cells, which means there are <u>two</u> of each <u>gene</u> — each gene can exist in <u>different forms</u> called alleles.

2) The different alleles cause different <u>versions</u> of a <u>characteristic</u> (e.g. blue eyes vs brown eyes) but only <u>one</u> version will actually show up. The allele whose version of the characteristic shows up is called the <u>dominant</u> allele. The other one is called <u>recessive</u>.

3) <u>Letters</u> are usually used to represent alleles in genetic diagrams. <u>Capital</u> letters are used for <u>dominant</u> alleles (e.g. 'C') and <u>recessive</u> alleles are shown with <u>small letters</u> (e.g. 'c').

4) For a <u>recessive characteristic</u> to be displayed <u>both</u> alleles must be <u>recessive</u> (e.g. cc). But to display a <u>dominant characteristic</u> the organism can be <u>either</u> CC or Cc, because the dominant allele <u>overrules</u> the recessive one.

Some Disorders have Genetic Causes

Some disorders are caused by <u>faulty genes</u>. Just as characteristics such as <u>hair colour</u> can be <u>inherited</u>, <u>genetic disorders</u> can also be <u>inherited</u>. Whether or not you are affected by the disorder depends on the type of allele you inherit — there are two good examples of this...

1) Cystic Fibrosis is Caused by a Recessive Allele

<u>Cystic fibrosis</u> affects the <u>cell membranes</u>. It causes the body to produce <u>thick sticky mucus</u> in the air passages and the digestive system.

1) The allele that causes cystic fibrosis is <u>recessive</u> (so we write this as 'f').

2) Because it's recessive, people with only <u>one</u> copy <u>won't</u> have the disorder — but will be <u>carriers</u>.

3) For someone to be a <u>sufferer</u>, <u>both</u> of their parents must be either <u>carriers or sufferers</u>.

Parents' characteristic: normal, but carrier   normal, but carrier
Parents' alleles: Ff   Ff
Gametes' alleles: F   f   F   f
Offspring's alleles: FF   Ff   Ff   ff
Offspring's characteristic: normal   carrier   carrier   sufferer

Both parents have one recessive allele. They'll be carriers but not sufferers — the dominant allele overrules the recessive one.

Half the gametes have the dominant allele and half have the recessive allele.

There's a 1 in 4 chance of a child having the disorder and a 2 in 4 (i.e. 50%) chance of a child being a carrier.

2) Huntington's is Caused by a Dominant Allele

<u>Huntington's disease</u> is a disorder of the <u>nervous system</u>. It results in <u>shaking</u>, <u>erratic body movements</u> and eventually severe <u>mental deterioration</u>.

1) The allele that causes it is <u>dominant</u> ('H').

2) This means that it can be inherited if just <u>one parent</u> carries the defective allele.

3) The "<u>carrier</u>" parent will also be a <u>sufferer</u>, but the symptoms don't start to appear until after the person is about 40. By this time the allele could already have been passed on to <u>children</u> and even to <u>grandchildren</u>.

carrier/sufferer   normal
Hh   hh
H   h   h   h
Hh   hh   Hh   hh
sufferer   normal   sufferer   normal

We know of over 4000 different genetic disorders...

This page is pretty nasty — there are some horrible words and a few <u>tricky ideas</u> to get your head around. <u>Alleles</u> are just <u>different versions</u> of the <u>same gene</u> — they can either be <u>recessive</u> or <u>dominant</u>.

Treating Genetic Disorders

As you've just seen, genetic disorders can be pretty nasty. But there may be a solution — there's been loads of <u>research</u> into the <u>causes</u> and the <u>treatment</u> of genetic diseases. Things like <u>gene therapy</u> and the <u>Human Genome Project</u> could mean genetic disorders will soon be a thing of the <u>past</u>. Fingers crossed.

Faulty Genes Make Faulty Proteins

1) Remember each <u>gene</u> codes for the production of a specific <u>protein</u> (see page 24).
2) If a gene is faulty, it might produce the <u>wrong</u> protein or <u>not produce</u> a protein at all.
3) This can cause <u>genetic disorders</u> such as cystic fibrosis and Huntington's disease (see page 26).

Gene Therapy Could be Used to Treat Genetic Disorders

1) <u>Gene therapy</u> is a new, <u>experimental treatment</u> for genetic disorders.
2) A <u>new</u>, <u>working</u> version of a faulty gene is inserted into a patient's cells.
3) These cells would then be able to make the <u>correct protein</u> and the <u>symptoms would disappear</u>.
4) Gene therapy would <u>target</u> areas badly affected by the disorder.
5) Scientists are currently trying to cure <u>cystic fibrosis</u> with gene therapy:

> One method being trialled uses a <u>virus</u> to insert a <u>healthy copy</u> of the gene into the cells in the <u>airways</u>. In the long term scientists hope to make these changes <u>permanent</u>, but so far trials testing gene therapy have only shown <u>very temporary</u> improvements.

6) Gene therapy still needs a lot of <u>research</u> and <u>testing</u>. <u>Potential dangers</u> of gene therapy have already been found, e.g. the introduced genetic material could disrupt other genes, causing a whole new set of medical problems.

The Faulty Genes Will Still be Passed On Though

1) Even if someone is treated with gene therapy their children might <u>still inherit</u> the <u>faulty gene</u>.
2) This is because gene therapy is <u>targeted</u> to the <u>area of the body</u> the disorder affects. The working, healthy gene is only incorporated into those specific cells and not into the <u>reproductive cells</u> that make eggs and sperm. It would be very, very difficult to target the reproductive cells using gene therapy.

Techniques Could Improve as Our Knowledge Increases

In the early 1990s scientists began working on the <u>Human Genome Project</u>. The big idea was to <u>find every single human gene</u>. Well... humans have around <u>25 000 genes</u>, they've <u>found most of them</u> — and now they're trying to figure out what each gene <u>does</u>. Knowing what each one does could have a massive effect on the <u>diagnosis</u> and <u>treatment</u> of genetic disorders...

1) <u>Identifying genes responsible for disease</u> — We don't know which genes cause <u>some</u> genetic disorders. If we did then it would mean that <u>gene therapy</u> might be possible for these disorders.
2) <u>Developing new medicines</u> — If we know which gene is <u>faulty</u> and what it does in the body then scientists might be able to design drugs that <u>reduce</u> the symptoms of a disorder.
3) <u>Accurate diagnosis</u> — Some genetic disorders are hard to <u>diagnose</u>, but if we know which faulty genes cause them, accurate testing would be a lot <u>easier</u>.

Gene therapy — talk things through with Dr Levi...

Gene therapy is a bit of an <u>ethical minefield</u>. It throws up all sorts of questions, including exactly <u>what</u> should be classed as a '<u>disorder</u>' that needs fixing. Many people would argue that treating <u>cystic fibrosis</u> is a <u>good thing</u>, but what about <u>high blood pressure</u>, or <u>ginger hair</u>... where does it stop?

Revision Summary for Section 2.2

Phew — finally, the end of that section. It was a bit of a tough one — lots of complex ideas, but hopefully it all made sense. Time to find out anyway. Have a go at these questions — if there are any you get wrong or can't answer just go back and reread the page, then come back and have another go.

1) Where in a cell would you find chromosomes?
2) What are genes?
3) Genes carry instructions — what are these instructions for?
4) What is an allele?
5) What are gametes?
6) How does gamete formation increase genetic variation?
7) How many chromosomes does a human gamete have?
8) Sexual reproduction is one way of increasing genetic variation. Give another.
9) State one example of a characteristic which is only affected by the environment (not by genes).
10) Describe how the following might be caused by both genes and the environment:
    a) Body mass,
    b) Height.
11) What is meant by the terms dominant allele and recessive allele?
12)* If the allele for blonde hair ('b') is recessive, which of the following will have blonde hair?
    a) BB,
    b) bb,
    c) Bb.
13) Sam has cystic fibrosis. Neither of his parents are sufferers. What does this tell you about his parents' alleles?
14) Sam's parents are expecting another child. What are the chances of it suffering from cystic fibrosis?
15) What is Huntington's disease?
16) For Huntington's to be inherited, how many parents need to have the defective allele?
17) Why do many people pass the Huntington's allele on to their children without realising they were carrying it?
18) How can a faulty gene lead to a genetic disorder?
19) How does gene therapy work?
20) How could gene therapy be used to treat cystic fibrosis?
21) Why is further research into gene therapy needed?
22) Even after gene therapy, the faulty genes might be passed on to the next generation. Why is this?
23) Roughly how many genes are there in the human genome?
24) Give three possible benefits of the Human Genome Project.

*See p.148 for answers.

Microorganisms and Disease

Microorganisms are organisms that are microscopic (unsurprisingly). There are 'good' ones and there are 'bad' ones, which cause horrid diseases that make you really ill. Here's a page about the bad ones...

Infectious Diseases are Caused by Pathogens

1) <u>Pathogens</u> are <u>microorganisms</u> that cause <u>disease</u>.

2) Human pathogens (like most microorganisms) can <u>reproduce very fast</u> inside the <u>body</u> — they love the <u>warm conditions</u>. They live off their host and give nothing in return.

3) Pathogens include some <u>bacteria</u>, some <u>fungi</u> and <u>all viruses</u>.
Here's a bit more about them:

1) Bacteria are Very Small Living Cells

1) Bacteria are <u>very small cells</u> (about 1/100th the size of your body cells) that can reproduce rapidly inside your body.

2) They can make you <u>feel ill</u> by doing <u>two</u> things:

    a) <u>damaging your cells</u>, b) <u>producing toxins</u> (poisons).

3) Some bacteria can cause diseases like <u>tuberculosis</u> (a lung disease) and <u>food poisoning</u>.

Bacteria are cells with no nucleus. The genes are free in the cell.

Some bacteria are <u>useful</u> if they're in the <u>right place</u>, like in your digestive system.

2) Viruses aren't Cells — They're Much Smaller

1) Viruses are <u>not cells</u>. They're <u>tiny</u>, about 1/100th the size of a bacterium. They're usually no more than a <u>coat of protein</u> around some genes.

2) They <u>replicate themselves</u> by <u>invading your cells</u> and using the <u>cell machinery</u> to produce many <u>copies</u> of themselves. The cell will usually then <u>burst</u>, releasing all the new viruses. This <u>cell damage</u> is what makes you feel ill.

3) Viruses can cause diseases like <u>measles</u>, <u>mumps</u>, <u>rubella</u>, <u>common colds</u> and <u>flu</u>.

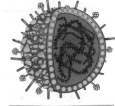

A horrid Flu Virus

3) Fungi are Living Cells

You don't need to know much about fungi but they do cause two common diseases — <u>athlete's foot</u> and <u>thrush</u>.

There are Two Main Ways Infectious Diseases can be Caught

The <u>more</u> of the pathogen that gets into the body, the <u>more likely</u> you are to get the disease.
There are <u>two</u> main ways in which diseases are <u>caught</u>:

1) Through <u>CONTACT</u> with an <u>INFECTED PERSON</u>. This can be through <u>direct</u> or <u>indirect</u> contact:

<u>DIRECT</u>, e.g. the <u>HIV virus</u> is spread through <u>body fluids</u>, and so can be passed on through sexual contact.

<u>INDIRECT</u>, e.g. <u>influenza</u> (<u>flu</u>) and the <u>common cold</u> viruses are spread through the <u>air</u> in sneezes and coughs.

2) Through <u>UNHYGIENIC CONDITIONS</u>. This can be from <u>contaminated food</u>, <u>water</u> or <u>surfaces</u>:

<u>FOOD</u>, e.g. <u>food poisoning</u> can be caused by badly prepared meat.

<u>WATER</u>, e.g. <u>cholera</u> is often caused by bacteria from sewage leaking into water supplies.

<u>SURFACES</u>, e.g. dirty surgical instruments or dirty food preparation surfaces.

Unhygienic conditions — dirty kitchens, students' bedrooms...

There are about 10 times more bacterial cells in your body than there are human cells... which is a pretty weird thought. Most of them are in your digestive system — they actually help your digestion. They're sometimes called 'good bacteria', while the nasty fellas on this page are labelled 'bad bacteria'.

Reducing the Spread of Disease

To reduce the spread of disease you need to reduce the chance of being <u>exposed</u> to pathogens. This means making things more hygienic and avoiding people with the sniffles...

There are Loads of Ways to Reduce the Spread of Disease

PRACTISING GOOD PERSONAL HYGIENE

1) <u>Washing your hands</u> — Washing your hands after you've been to the toilet, handled raw meat or been digging for worms <u>reduces the chance</u> of nasty pathogens that may have got on your hands being <u>transferred to your mouth</u>, <u>eyes</u> or <u>nose</u> (where they can get into the body and cause disease).

2) <u>Protective clothing</u> — Covering up when in close contact with pathogens, e.g. <u>wearing rubber gloves</u> while cleaning the toilet (which I'm sure you do all the time), helps to stop pathogens getting from your hands into your body.

STERILISING EQUIPMENT

1) <u>Using high temperatures</u> — Doctors often use high temperatures to kill microorganisms on <u>surgical equipment</u> so the patient doesn't get infected.

2) <u>Using radiation</u> — Some things can't be sterilised using heat as they would melt, e.g. plastic surgical equipment like breathing tubes. They're often bombarded with <u>gamma rays</u> (a type of radiation, see p.33) instead. The radiation <u>kills</u> all the microorganisms on the equipment.

DISINFECTING SURFACES

1) Disinfectants are <u>chemicals</u> that kill microorganisms. They include things like <u>bleach</u> and <u>chlorine</u>.

2) They're used to kill microorganisms on <u>surfaces</u>, e.g. on lab benches, kitchen worktops and hospital ward floors. This means there are <u>fewer pathogens</u> that can be transferred from the surface to the body.

3) Disinfectants are normally pretty <u>toxic</u> so you can't use them to disinfect skin.

DISINFECTING SKIN USING ANTISEPTICS

1) <u>Using antiseptics</u> — Antiseptics are also <u>chemicals</u> that kill microorganisms. They include things like <u>alcohol</u> and <u>iodine</u>.

2) They're <u>less toxic</u> than disinfectants so they're used to kill microorganisms on <u>skin</u>.

3) Alcohol is often used to clean the skin <u>before injections</u> and iodine is used to <u>clean cuts</u> and to clean the skin <u>before surgery</u> (it makes the skin orange).

OTHER WAYS

1) <u>Keeping kitchens hygienic</u> — If kitchen surfaces are clean there are fewer pathogens that can get onto food that you eat.

2) <u>Cooking food properly</u> — This kills any pathogens that may be present in food so you're less likely to get food poisoning.

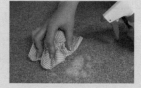

3) <u>Avoiding unprotected sex</u> — This reduces the chance of contracting sexually transmitted infections like HIV and herpes. Not having sex is the only sure-fire way of prevention.

4) <u>Avoiding ill people</u> — Avoiding people with a cold or flu will decrease your chances of catching it. People with very contagious infectious are sometimes quarantined in hospital for the same reason.

There are more bacteria on the kitchen sink than the toilet seat...

For your exam you need to <u>learn</u> the ways to reduce the spread of disease... it's pretty handy knowledge for your own life too. <u>Cleanliness</u> is the key. Look at your hands, they're filthy...

The Body's Defence Systems

Even if you're the <u>most hygienic person</u> on the planet you'll still end up being <u>exposed to microorganisms</u>. But <u>not to worry</u> because your body has <u>two main lines of defence</u> against them — 1) <u>stopping</u> the little nasties from <u>getting in</u>, and 2) <u>destroying them</u> if they do manage to get past the first line of defences.

Skin, Hairs and Mucus Stop Microorganisms Getting In

1) Your <u>skin</u>, plus <u>hairs</u> and <u>mucus</u> in your respiratory tract (breathing pipework), stop a lot of nasties getting inside your body.
2) Small fragments of cells (called <u>platelets</u>) help your <u>blood clot quickly</u> to seal wounds. This prevents microorganisms getting into the body through <u>cuts</u>.

The Immune System Attacks Microorganisms that Do Get In

If something does make it through the first defences your <u>immune system</u> kicks in. The most important part of your immune system is the <u>white blood cells</u>. They travel around in your blood and crawl into every part of you, constantly patrolling for microbes. When they come across an invading microbe they have three lines of attack:

1. Consuming Them

White blood cells can <u>engulf</u> foreign cells and <u>digest</u> them.

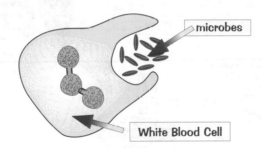

2. Producing Antibodies

1) Every invading pathogen has <u>unique molecules</u> (called <u>antigens</u>) on its surface.
2) When your white blood cells come across a <u>foreign antigen</u> (i.e. one it doesn't recognise), they'll start to produce molecules called <u>antibodies</u> to lock on to and <u>kill</u> the invading pathogens. The antibodies produced are <u>specific</u> to that type of antigen — they won't lock on to any others.
3) Antibodies are then produced <u>rapidly</u> and flow all round the body to <u>kill all similar bacteria or viruses</u>.
4) If the person is infected with the <u>same pathogen</u> again the white blood cells will rapidly produce the antibodies to kill it — the person is <u>naturally immune</u> to that pathogen and will be able to <u>fight it off</u>.

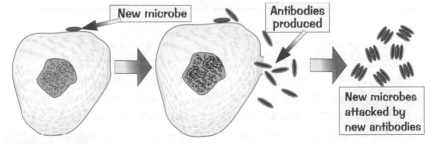

3. Producing Antitoxins

These are natural antidotes to the toxins produced by the <u>invading pathogens</u>.

Fight disease — blow your nose with boxing gloves...

So by now you might have worked out that if you have a <u>low</u> level of white blood cells you'll be more susceptible to <u>infections</u>. In fact, HIV/AIDS doesn't kill people <u>directly</u> — it just makes it easier for something else to by <u>attacking</u> white blood cells and <u>weakening</u> the immune system. However, other diseases (e.g. leukaemia) can <u>increase</u> the number of white blood cells — and that's no good either.

Vaccination

The body is pretty good at fending off pathogens — but some infections can be pretty serious and can kill you. Scientists have developed a way to <u>protect</u> against some of these infections — it's called <u>vaccination</u>. This means we don't always have to deal with the problem once it's happened — we can <u>prevent</u> it happening in the first place. Grand.

Vaccinations Help to Prevent Disease

1) When you're infected with a <u>new</u> microorganism it can take your white blood cells a while to produce the antibodies to deal with it. In that time you can get <u>very ill</u>, or maybe even die.

2) To avoid this you can be <u>vaccinated</u> against some diseases, e.g. <u>polio</u>, <u>tuberculosis</u>, <u>measles</u>, <u>mumps</u> and <u>rubella</u>.

3) Here's how it works:

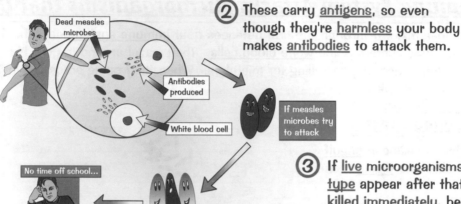

(1) Vaccination involves injecting <u>dead or inactive</u> microorganisms into the body.

Dead measles microbes

Antibodies produced

White blood cell

No time off school...

They are recognised quickly and attacked

Antibodies

(2) These carry <u>antigens</u>, so even though they're <u>harmless</u> your body makes <u>antibodies</u> to attack them.

If measles microbes try to attack

(3) If <u>live</u> microorganisms of the <u>same type</u> appear after that, they'll be <u>killed immediately</u>, because your body can produce the right antibodies much more quickly the second time around. Cool.

1) Vaccines have helped <u>control</u> lots of infectious diseases that were once <u>common</u> in the UK (e.g. polio, measles, whooping cough, rubella, mumps, TB, tetanus...).

2) And if an outbreak does occur, vaccines can <u>slow down</u> or <u>stop</u> the spread (if people don't catch the disease, they won't pass it on).

3) Vaccination is now used all over the world. <u>Smallpox</u> no longer occurs at all, and <u>polio</u> infections have fallen by 99%.

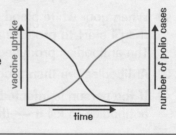

Vaccines are Very Safe — But There is Some Risk

1) There are disadvantages to being vaccinated — sometimes it doesn't give you <u>immunity</u>, and sometimes you can have a <u>bad reaction</u> (e.g. swelling, or maybe something more serious such as a fever or seizures). But bad reactions are very <u>rare</u> — vaccines are very <u>safe</u>.

2) Deciding whether to have a vaccination means balancing risks — the risk of <u>catching</u> the disease if you <u>don't</u> have a vaccine, against the risk of having a <u>bad reaction</u> if you <u>do</u>. As always, you need to look at the <u>evidence</u>. For example, if you get <u>measles</u> (the <u>disease</u>), there's about a <u>1 in 15</u> chance that you'll get <u>complications</u> (e.g. pneumonia) — and about 1 in 1000 people who get measles actually <u>die</u>. However, the number of people who have a problem with the <u>vaccine</u> is more like <u>1 in 100 000</u>.

Prevention is better than cure...

In the exam you might be asked to <u>interpret</u> a graph of vaccine uptake and disease occurrence. Don't panic — just take each line on the graph in turn and <u>explain what happens to it</u>, e.g. whether it goes up, or down or stays level. You might also have to interpret <u>data</u> — it's the <u>same deal</u>, just look for the <u>trends</u>.

Effects of Radiation on the Body

Radiation can be used to <u>detect</u> medical disorders and to <u>treat</u> diseases.
But before you get into the uses of radiation, you need to know what it is.

Nuclear Radiation: Alpha, Beta and Gamma (α, β and γ)

1) There are <u>three</u> different types of nuclear radiation — <u>alpha</u> (α), <u>beta</u> (β) and <u>gamma</u> (γ).

2) They all come from <u>radioactive sources</u>, e.g. uranium.

3) Nuclear radiation causes <u>ionisation</u> (see below), so it's sometimes called <u>ionising radiation</u>.

RADIATION DAMAGES LIVING CELLS

1) Nuclear radiation can <u>bash into molecules</u> inside living <u>cells</u>, knocking bits off them.

2) This is called <u>ionisation</u> — and it can <u>damage</u> or <u>destroy</u> the cell.

3) <u>Lower doses</u> tend to cause <u>minor</u> damage without killing the cell. This can make the cells <u>divide</u> out of control, causing <u>cancer</u>.

4) Higher doses tend to kill the cells completely. If a lot of body cells get blatted at once, this causes <u>radiation sickness</u> — nausea, a weakened immune system, hair loss, etc.

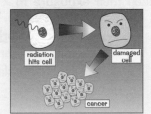

The Three Types of Radiation Have Different Properties

Alpha Particles

1) They're relatively <u>big</u>, <u>heavy</u> and <u>slow moving</u>.

2) They <u>can't pass through your skin</u> to get into your body.

3) You can have <u>problems</u> if they get inside your body some other way though, e.g. if you <u>swallow</u> or <u>breathe in</u> a source of alpha particles. They're <u>very ionising</u>, so they do <u>lots</u> of damage in a very <u>localised area</u>.

Beta Particles

1) They're <u>small</u>, <u>light</u> and <u>fast moving</u>.

2) They can be stopped by a <u>thin sheet of aluminium</u> (or something similar), but they pass through your skin <u>fairly easily</u>. So a beta source <u>outside</u> the body can still cause damage to sensitive internal organs.

3) They're not as ionising as alpha though, so they tend to cause <u>less damage</u>.

Gamma Rays

1) They're different from the other two types of radiation in that they're actually high-energy <u>electromagnetic waves</u> (see page 103).

2) They're <u>very fast</u>, and can go a <u>long way</u>. It takes <u>thick lead</u> or <u>concrete</u> to stop a gamma ray.

3) They're only <u>weakly ionising</u> — so they'll tend to pass straight through your body without doing much damage.

<u>X-RAYS</u> are also a type of radiation. They're very <u>similar</u> to <u>gamma</u> rays. The only real difference is that X-rays don't come from radioactive sources — they're made in a different way.

I once beta particle — it cried for ages...

The bigger the dose of radiation the more likely your cells are to be damaged. Sadly, scientists know all this stuff because of studies on radiation disasters like Chernobyl (a nuclear power station that blew up).

Medical Uses of Radiation

A radiographer is a medical professional who uses radiation. They can help work out what's wrong with a patient, e.g. by taking X-ray pictures. They can also treat some cancers using radiotherapy.

X-Rays are Used in Hospitals to Detect Broken Bones

1) Radiographers in hospitals take X-ray photographs of people to see whether they have broken bones.
2) X-rays pass easily through flesh but not through denser material like bones or metal.
3) X-rays can cause cancer, so radiographers wear lead aprons and stand behind a lead screen or leave the room to keep their exposure to X-rays to a minimum.

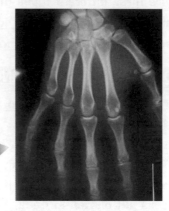

The brighter bits are where fewer X-rays get through. This is a negative image. The plate starts off all white.

Radiotherapy — the Treatment of Cancer Using Gamma Rays

1) Radiation can cause cancer, but it can also be used to treat it.
2) High doses of gamma rays will kill all living cells, including cancerous ones.
3) The gamma rays have to be directed carefully and at just the right dosage so as to kill the cancer cells without damaging too many normal cells.

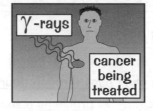

4) Like with X-rays, the radiographer has to be protected from the radiation.
5) However, a fair bit of damage is inevitably done to normal cells, which makes the patient feel very ill. But if the cancer is successfully killed off in the end, then it's usually worth it.

Gamma Rays are Also Used to Sterilise Surgical Instruments

1) Medical instruments can be sterilised by exposing them to a high dose of gamma rays, which will kill all microbes.

See p.30 for more on how this helps reduce the spread of disease.

2) The great advantage of radiation over boiling is that it doesn't involve high temperatures, so heat-sensitive things like thermometers and plastic instruments can be totally sterilised without damaging them.

Radiographers are like teachers — they can see right through you...

X-rays can be used for other things apart from detecting broken bones — e.g. they can be used to look for bullets after people have been shot. And if you accidentally swallow something you shouldn't have, e.g. a Terry Wogan action figure, you'll have an X-ray to make sure it's not stuck somewhere dangerous.

Section 2.3 — The Body at Risk

Health & Medicine

Healthy Diet

Not eating the right things and not getting enough exercise can be pretty bad for you...

A Healthy Diet Has to be Balanced

1) You need to eat the right amount of food for your body size and level of activity.

2) But that's not all. Different food groups have different uses in the body, so you need to have the right balance of foods as well.
   You need:

   ...enough carbohydrates (sugars) and fats to keep warm and provide energy,
   ...enough protein for growth, cell repair and cell replacement,
   ...enough fibre to keep everything moving smoothly through your digestive system,
   ...and tiny amounts of various vitamins and minerals to keep your skin, bones, blood and everything else generally healthy.

Some foods are generally thought of as healthy, and some foods are generally thought of as unhealthy if you eat TOO MUCH of them.

This doesn't mean that you should avoid saturated fat, salt and sugar altogether. The body needs these things in small amounts to function properly.

Generally healthy foods	Generally unhealthy if eaten in EXCESS
Fresh fruit (contain lots of minerals and vitamins)	Saturated fat — found in foods like pies and biscuits
Fresh vegetables (contain lots of minerals and vitamins)	Salt — found in high levels in some ready meals
	Sugar — found in high levels in some fizzy drinks and chocolate

This also doesn't mean you should only eat, say, oranges all day. Different fruits and vegetables contain different vitamins and minerals, for example green vegetables and citrus fruits contain lots of vitamin C and dark green leafy vegetables contain lots of iron. That's why it's important to eat a mixture of different fruits and veg.

An Unhealthy Diet and Lack of Exercise can Cause Disease

1) People who do little exercise and eat an unhealthy diet that contains lots of fat or sugar can become overweight or obese.

2) People who are overweight are more likely to suffer from certain diseases, especially when they're older:

DIABETES

Diabetes is a disorder where the person's body can't use glucose properly.

HEART DISEASE

If a person is overweight their heart has to work much harder to pump the blood around their body. Also fat might clog up their arteries so the heart has to work even harder to squeeze the blood through their narrow arteries. This can cause high blood pressure and heart failure.

Why is it that all the tasty foods are bad for you?

It all comes down to how much of each type of food you eat. Too much of one thing can cause obesity, which puts you at risk of a whole load of disorders. Too much of other things can also be bad for you, e.g. too much vitamin A can kill you (but you'd have to eat thousands of carrots to get too much).

Use of Drugs to Treat Disease

Vaccinations, living hygienically and choosing a healthy diet can help prevent disease — but once you get a disease you need to be able to treat it. Drugs are constantly being developed to treat diseases.

Drugs Can be Beneficial or Harmful

1) Drugs are substances that alter the way the body works. Some drugs are medically useful, such as antibiotics (e.g. penicillin) or anti-inflammatories (e.g. aspirin). But many drugs are dangerous if misused (see p.38-39).

2) This is why you can buy some drugs over the counter at a pharmacy, but others are restricted so you can only get them on prescription — your doctor decides if you should have them.

Some Drugs Just Relieve Symptoms — Others Cure the Problem

1) Painkillers (e.g. aspirin and paracetamol) are drugs that relieve pain (no, really). However, they don't actually tackle the cause of the pain (the cause is disease) — they just help to reduce the symptoms.

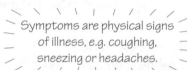
Symptoms are physical signs of illness, e.g. coughing, sneezing or headaches.

2) Other drugs do a similar kind of thing — reduce the symptoms without tackling the underlying cause. For example, lots of "cold remedies" don't actually cure colds.

3) Antibiotics (e.g. penicillin) work differently — they actually kill (or harm) the bacteria causing the problem without killing your own body cells.

4) However, antibiotics don't destroy viruses. Viruses reproduce using your own body cells, which makes it very difficult to develop drugs that destroy just the virus without killing the body's cells.

5) Flu and colds are caused by viruses. Usually you just have to wait for your body to deal with the virus, and relieve the symptoms if you start to feel really grotty. There are some antiviral drugs available, but they're usually reserved for very serious viral illnesses (such as AIDS and hepatitis).

Bacteria Can Become Resistant to Antibiotics

1) Antibiotics were an incredibly important (but accidental) discovery. Some killer diseases (e.g. pneumonia and tuberculosis) suddenly became much easier to treat. The 1940s are sometimes called the era of the antibiotics revolution — it was that big a deal.

2) Unfortunately, bacteria evolve (adapt to their environment). If antibiotics are taken to deal with an infection but not all the bacteria are killed, those that survive may be resistant to the antibiotic and go on to flourish. This process leaves you with an antibiotic-resistant strain of bacteria — not ideal.

3) A good example of antibiotic-resistant bacteria is MRSA (methicillin-resistant *Staphylococcus aureus*) — it's resistant to the powerful antibiotic methicillin.

4) This is why it's important for patients to always finish a course of antibiotics, and for doctors to avoid over-prescribing them.

Antibiotic resistance is inevitable...

Antibiotic resistance is scary. Bacteria reproduce quickly, and so are pretty fast at evolving to deal with threats (e.g. antibiotics). If we were back in the situation where we had no way to treat bacterial infections, we'd have a nightmare. So do your bit, and finish your course of antibiotics.

Testing New Medical Drugs

Scientists at pharmaceutical companies can't just develop a drug and sell it to the general public straight away — it has to be <u>tested</u> first to make sure it works and that it's <u>safe</u>.

Medical Drugs are Developed and Then Thoroughly Tested

This is what usually happens...

1 <u>Computer models</u> are often used in the early stages of drug development — these <u>simulate</u> a human's response to a drug. This can <u>identify promising drugs</u> to be tested in the next stage (but sometimes it's not as accurate as actually seeing the effect on a <u>live organism</u>).

2 Drugs are then developed further by testing on <u>human tissues</u> in the lab. However, you can't use human tissue to test drugs that affect <u>whole</u> or <u>multiple</u> body systems, e.g. testing a drug for blood pressure must be done on a whole animal with an intact circulatory system.

3 The next step is to develop and test the drug using <u>live animals</u>. The law in Britain states that any new drug must be tested on <u>two</u> different <u>live mammals</u>.

4 After the drug has been tested on animals it's tested on <u>healthy human volunteers</u> in a small <u>clinical trial</u> — this should determine whether there are any <u>side effects</u>. If the drug doesn't have too many side effects it's then tested on a <u>small</u> number of <u>patients</u>, and then a <u>large</u> number of patients. At each stage a '<u>control group</u>' is given an identical-looking drug (called a <u>placebo</u>). This is done so scientists can see if the drug is having an effect and the patients don't just feel better because they <u>think</u> they're being treated.

There are Issues Surrounding Drug Testing

Here are a few arguments for and against testing drugs on animals:

1) Some people think that testing drugs on animals is <u>cruel</u>.

2) Others believe this is the <u>safest</u> way to make sure a drug <u>isn't dangerous</u> before it's given to humans.

3) Some people think that animals are <u>so different</u> from humans that testing on animals is <u>pointless</u>.

4) Other people think that mammals have <u>very similar</u> body systems and genes to humans (especially primates, e.g. monkeys) so they are a <u>good model</u> of how a drug might work in humans.

5) Some people believe that the suffering caused to animals <u>doesn't outweigh the benefits</u> of testing the drug on animals. Other people think that it <u>does</u>.

The UK animal testing regulations are some of the strictest in the world. When doing <u>research</u>, scientists are required to <u>replace</u> animal testing with another method if possible, <u>reduce</u> the number of animals used as much as they can and <u>refine</u> the experimental techniques used, to minimise pain. However, to test <u>finished drugs</u> animals must be used — it's the <u>law</u>.

There are issues surrounding clinical trials too. Some people think it's unethical to put <u>human lives at risk</u> even if there's a potential benefit. Some people think clinical trials are too long (they can take years) and think it's unethical to <u>withhold potentially life-saving drugs</u>.

This page might be a bit testing but you need to know it...

Testing drugs on animals is a pretty controversial topic — but whatever your views are on it you need to know <u>both sides</u> of the story. Also, don't get it mixed up with testing cosmetics on animals.

Recreational Drugs

Some drugs are also used <u>recreationally</u> (i.e. for fun). Some of these are <u>legal</u>, others <u>illegal</u>. They can all cause harm to the body, but some are more <u>harmful</u> than others.

Some Recreational Drugs are Legal...

1) ALCOHOL — Alcohol is found in <u>alcoholic drinks</u>. It affects the <u>nervous system</u> and slows down the body's reactions (see the next page for more).

2) NICOTINE — This is the <u>addictive</u> drug found in <u>cigarettes</u>. It's a <u>stimulant</u> — it increases the activity of the brain and makes you feel <u>more alert</u>. See the next page for more.

...and Some are Illegal (without a prescription from your doctor)

1) ANTIDEPRESSANTS — These are <u>medical</u> drugs that are designed to <u>reduce depression</u>. These are <u>stimulant</u> drugs. You can only get them if your <u>doctor</u> prescribes you the drugs. Some people take antidepressants illegally.

2) AMPHETAMINES — Also known as <u>speed</u>. These are <u>stimulant</u> drugs. They make you feel <u>more alert</u>.

3) COCAINE — This is a <u>stimulant</u> — it revs up the nervous system and makes you feel more <u>alert</u>.

4) BARBITURATES — These are medical drugs that are <u>used as sedatives</u> and to help you <u>sleep</u>. You can only get them from your doctor. They can make you feel like you're <u>drunk</u>.

5) HEROIN — A form of heroin is used in <u>hospitals</u> as a <u>painkiller</u>. But people also use heroin illegally because it makes them '<u>high</u>'.

Lots of Drugs Can Be Addictive and All Can Be Harmful

Some people become <u>addicted</u> to drugs — this means they have a physical need for the drug. This is because the drug changes some of the <u>chemical processes</u> in the body. If they don't get the drug they get <u>withdrawal symptoms</u>, e.g. nicotine withdrawal when quitting smoking causes irritability and headaches.

The misuse of drugs can cause <u>harmful physical</u> and <u>psychological</u> effects. A few examples are:

1) <u>Alcohol</u> in excess causes <u>liver failure</u> and <u>brain damage</u>.
2) <u>Nicotine</u> causes <u>high blood pressure</u>.
3) <u>Cocaine</u> causes an increased risk of <u>heart attack</u>.
4) <u>Stimulants</u> can cause <u>insomnia</u> (when you can't sleep).
5) Sharing needles in <u>heroin</u> use increases the risk of contracting <u>HIV</u>.

Ultimately an <u>overdose can kill</u>. This is caused by too much of the drug in your body, which can trigger things like heart attacks, strokes (a blood clot in the brain) and respiratory failure (where you just stop breathing). Scary.

Learn all about drugs — and then forget them...

Both legal and illegal drugs can have major impacts on your body, some <u>permanently</u>. The <u>extent of the damage</u> usually matches the <u>dose</u> of the drug — the more you take (and the more often you take it) the more likely you are to damage your body.

Recreational Drugs

You might think that just because alcohol and tobacco are legal that they don't do you much harm. Well, think again. They <u>can</u> cause serious harm and you need to know all about them...

Drinking Alcohol Can Damage the Liver and Brain

1) The main effect of alcohol is to <u>reduce the activity</u> of the <u>nervous system</u> — slowing your reactions.

2) It can also make you feel <u>less inhibited</u> — which can help people to socialise and relax with each other.

3) However, too much leads to <u>impaired judgement</u>, <u>poor balance</u> and <u>coordination</u>, <u>lack of self-control</u>, <u>unconsciousness</u> and even <u>coma</u>.

4) Alcohol in excess also causes <u>dehydration</u>. It can damage <u>brain cells</u>, causing a noticeable <u>drop</u> in <u>brain function</u>. And too much drinking causes <u>severe damage</u> to the <u>liver</u>, leading to <u>liver disease</u>.

5) There are <u>social</u> costs too. Alcohol is linked with way more than half of <u>murders</u>, <u>stabbings</u> and <u>domestic assaults</u>. And alcohol misuse is also a factor in loads of divorces and cases of child abuse.

Smoking Tobacco Can Cause Quite a Few Problems Too

It affects the Circulatory System...

1) Tobacco smoke contains <u>carbon monoxide</u> — this <u>combines</u> with haemoglobin (see p.13) in red blood cells, meaning the blood can carry <u>less oxygen</u>. In pregnant women, this can deprive the <u>fetus</u> of oxygen, leading to the baby being born <u>underweight</u>.

2) Smoking also causes <u>disease</u> of the <u>heart</u> and <u>blood vessels</u> (leading to <u>heart attacks</u> and <u>strokes</u>).

...and the Respiratory System

1) Cigarette smoke damages the <u>lungs</u> (leading to diseases like <u>emphysema</u> and <u>bronchitis</u>).

2) <u>Tobacco smoke</u> contains <u>carcinogens</u> — chemicals that can lead to <u>cancer</u>. Lung cancer is <u>way more common</u> among <u>smokers</u> than non-smokers (see below).

3) And the <u>tar</u> in cigarettes damages the <u>cilia</u> (little hairs) in your lungs and windpipe. These hairs, along with <u>mucus</u>, catch a load of <u>dust</u> and <u>microbes</u> before they reach the lungs. When these cilia are damaged, it's harder for your body to eject stuff that shouldn't be there, which makes <u>chest infections</u> more likely.

And to top it all off, smoking tobacco is <u>addictive</u> — due to the <u>nicotine</u> in tobacco smoke.

Smoking and Lung Cancer are Now Known to be Linked

1) In the first half of the 20th century it was noticed that <u>lung cancer</u> and the popularity of <u>smoking</u> increased <u>together</u>. And studies found that far more <u>smokers</u> than <u>non-smokers</u> got lung cancer.

2) But it didn't directly <u>prove</u> that smoking <u>caused</u> lung cancer. Some people (especially in the tobacco industry) argued that there was some <u>other</u> factor (e.g. a person's <u>genes</u>) that caused lung cancer and made people more likely to smoke.

3) Later research eventually <u>disproved</u> these claims. Now even the tobacco industry has had to admit that smoking does <u>increase</u> the <u>risk</u> of lung cancer.

People have been smoking tobacco for thousands of years...

It's the <u>nicotine</u> in cigarettes that is addictive. Smokers who want to <u>quit</u> can use <u>patches</u> or <u>gum</u> that contain nicotine. This helps them to reduce their withdrawal symptoms whilst they try to <u>break the habit</u> of smoking. They then gently wean themselves off the nicotine patches or gum. Genius.

Revision Summary for Section 2.3

That was a pretty long section, but a pretty interesting one (or at least I thought it was interesting — maybe I need a hobby). It covered everything from nasty pathogens and how your body deals with them to how to prevent and treat disease... phew. Well, no rest for the wicked, so here's some questions to make sure all that info didn't just get stored in the 'read it and forget it' part of your brain.

1) What are pathogens?
2) How do bacterial pathogens make you feel ill?
3) Give two examples of diseases caused by: a) bacteria, b) viruses, c) fungi.
4) What are the two main ways infectious diseases can be caught?
5) How does washing your hands reduce your chances of getting ill?
6) Give two ways doctors might sterilise surgical equipment.
7) What are disinfectants? Give two examples of where you might use them.
8) In what situation might an antiseptic be used?
9) What might you do in your kitchen to reduce the chances of catching food poisoning?
10) Name four things that stop microorganisms from getting into the body.
11) How do platelets help prevent microorganisms getting into the body?
12) What are the three main ways that white blood cells defend the body against microorganisms?
13) Name five diseases you can be vaccinated against.
14) Explain how vaccination works.
15) Name two diseases that vaccination has almost wiped out.
16) Give two disadvantages of vaccinations.
17) What are the three different types of nuclear radiation?
18) How does ionising radiation damage cells? What disease can this cause?
19) What will stop: a) an alpha particle, b) a beta particle, c) a gamma ray?
20) What are X-rays used to detect?
21) Why do radiographers wear lead aprons when they take X-rays?
22) Describe how radiotherapy works. Why does the radiation have to be carefully targeted?
23) Give one other medical use for radiation.
24) Name five things that you need in your diet to stay healthy.
25) Which types of food are generally unhealthy if you eat too much of them?
26) Name two diseases that overweight people are more likely to suffer from.
27) Give an example of a type of drug that relieves symptoms of a disease.
28) Give an example of a type of drug that treats the cause of a disease.
29) Which microorganisms can antibiotics not kill?
30) Give one problem with using antibiotics a lot.
31) Why are drugs tested before they are sold?
32) Describe the four stages a drug goes through during its development and testing.
33) Write a balanced argument for and against testing drugs on animals.
34) Name two legal and five illegal recreational drugs.
35) What is drug addiction?
36) Give an example of how a drug harms the body.
37) Give two ways in which alcohol damages the body.
38) How does smoking affect: a) the circulatory system, b) the respiratory system?

Specialised Plant Cells

Agriculture and farming are both really important — without them there wouldn't be much to eat. Plants play a pretty major role, as you'll see over the next few pages. The reason plants are so great is that they make food using sunlight — but to do this they need some specialist cells...

Plant Cells are Like Animal Cells... with a Few Extras:

Plants have three things in common with animal cells:

1) Nucleus contains genetic material that controls what the cell does.

2) Cytoplasm contains enzymes that speed up biological reactions.

3) Cell membrane holds the cell together and controls what goes in and out.

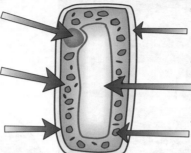

Three extras that only plant cells have:

1) Rigid cell wall made of cellulose gives support for the cell.

2) Large vacuole contains cell sap — a weak solution of sugar and salts.

3) Chloroplasts containing chlorophyll for photosynthesis. Found in the green parts of plants.

Some Plant Cells are Specialised for Their Job:

See next page for more on photosynthesis.

1) Palisade Leaf Cells are Adapted for Photosynthesis

1) They're packed with chloroplasts for photosynthesis.
2) Their tall shape means a large surface area is exposed for absorbing carbon dioxide from the air in the leaf.
3) The tall shape also means there's a good chance of light hitting a chloroplast before it reaches the bottom of the cell.

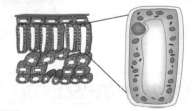

2) Guard Cells are Adapted to Open and Close

1) The underside of leaves are full of little holes called stomata. They're there to let gases like carbon dioxide and oxygen in and out for photosynthesis. Stomata are surrounded by special cells called guard cells.
2) Guard cells have a special kidney shape which opens and closes the stomata as the cells go turgid (plump) or flaccid (limp).
3) They have thin outer walls and thickened inner walls which make this opening and closing function work properly.
4) They're also sensitive to light — they close at night to conserve water without losing out on photosynthesis.

3) Root Hair Cells are Adapted to Take Water and Minerals from the Ground

1) The cells on plant roots grow into long 'hairs' that stick out into the soil.
2) This gives the plant a big surface area for absorbing water and minerals from the soil.

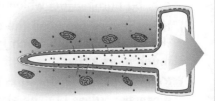

Plants are specialised to work in prisons — they guard cells...

It might not be the most exciting way to start a new section, but you're still going to have to learn it all. Specialised plant cells are specialised for a reason — to carry out a specific job. The examiner will love it if you know the types of specialised cells, the jobs they carry out and what makes them so darned special.

What Plants Need

Plants need food. <u>Photosynthesis</u> is the process that produces 'food' in plants. The 'food' it produces is <u>glucose</u>. Photosynthesis takes place in the <u>leaves</u> of all <u>green plants</u> — this is what leaves <u>are for</u>.

Photosynthesis Makes Glucose Using Sunlight

<u>Four</u> things are <u>needed</u> for photosynthesis to happen:

1) <u>CARBON DIOXIDE</u> — enters the leaf from the surrounding <u>air</u>.

2) <u>WATER</u> — comes from the <u>soil</u>, up the stem and to the <u>leaves</u>.

3) <u>LIGHT</u> — usually from the <u>Sun</u>, it provides <u>energy</u> for the process.

4) <u>CHLOROPHYLL</u> — the <u>green</u> <u>substance</u> which is found in <u>chloroplasts</u> and which makes leaves look <u>green</u>. Chlorophyll absorbs the <u>energy</u> in <u>sunlight</u> and uses it to combine <u>carbon dioxide</u> and <u>water</u> to produce <u>glucose</u>. Oxygen is simply a by-product.

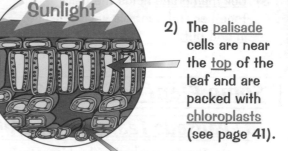

Sunlight

<u>Leaves</u> have <u>three</u> key features which make them good at photosynthesis:

1) They're <u>thin</u> and <u>flat</u> — this provides a big <u>surface area</u> to catch <u>lots</u> of sunlight.

2) The <u>palisade</u> cells are near the <u>top</u> of the leaf and are packed with <u>chloroplasts</u> (see page 41).

3) <u>Guard cells</u> control the movement of gases into and out of the leaf.

Photosynthesis can be Written as an Equation:

Carbon dioxide + Water $\xrightarrow[\text{chlorophyll}]{\text{SUNLIGHT}}$ Glucose + Oxygen

You can Artificially Create the Ideal Conditions for Photosynthesis

The problem is, some of the things needed aren't always there in the <u>right amounts</u>. Photosynthesis can be helped along by artificially creating the ideal conditions in <u>glass houses</u> (big greenhouses to you and me).

1) <u>Light</u> is always needed for photosynthesis, so <u>commercial farmers</u> often supply <u>artificial light</u> after the Sun goes down to give their plants <u>more</u> quality <u>photosynthesis time</u>.

2) Farmers can <u>increase</u> the level of <u>carbon dioxide</u> in the greenhouse, e.g. by <u>burning fuel</u> which makes carbon dioxide as a by-product.

3) If the farmer can keep the conditions <u>just right</u> for photosynthesis, the plants will grow much <u>faster</u> and a <u>decent crop</u> can be harvested much more <u>often</u>.

4) Keeping plants <u>enclosed</u> in a glass house also makes it easier to keep them free from <u>pests</u> and <u>diseases</u>. Glass houses also <u>trap</u> the Sun's <u>heat</u> so that it doesn't get too <u>cold for the plants</u>.

Live and Learn...

It's not just poor GCSE students who need to learn about how plants <u>photosynthesise</u>. Farmers use this information to get the <u>most</u> out of their crops. Those that got stuck in at school know how to grow things like <u>lettuce</u>, <u>tomatoes</u> and even <u>strawberries</u> all year round — even in the UK's cold, dark, rubbish climate.

What Plants Need

Plants are important in <u>nutrient cycles</u>. They take <u>minerals</u> from the soil so they can grow <u>big</u> and <u>healthy</u>, and then, after all that hard work, we eat them.

Plants Need Minerals for Healthy Growth

Plants need certain <u>elements</u> so they can produce important compounds. They get these elements from <u>minerals</u> in the <u>soil</u>. The minerals are usually present in the <u>soil</u> in quite <u>low concentrations</u> — if there aren't enough of them, plants suffer <u>deficiency symptoms</u>.

1) *Nitrates*

— these are needed to make proteins, which are needed for <u>cell growth</u>. If a plant can't get enough nitrates it will be <u>stunted</u> and will have <u>yellow older leaves</u>.

2) *Phosphates*

— they're needed for <u>respiration</u> and <u>growth</u>. Plants without enough phosphate have <u>poor root growth</u> and <u>purple older leaves</u>.

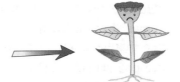

3) *Potassium*

— to help the <u>enzymes</u>. If there's not enough potassium in the soil, plants have <u>poor flower and fruit growth</u> and <u>discoloured leaves</u>.

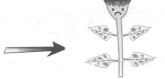

4) *Magnesium (in smaller amounts)*

— this is essential for making <u>chlorophyll</u> (needed for <u>photosynthesis</u>). Plants without enough magnesium have <u>yellow leaves</u>.

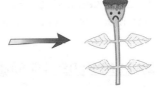

Fertilisers Contain the Main Nutrients

Gardeners and farmers often add <u>fertilisers</u> to make sure plants get all the minerals they need.

1) Fertilisers can either be <u>organic</u> (e.g. manure) or <u>inorganic</u> (artificial).

2) Fertilisers usually contain a mixture of the <u>main minerals</u> (<u>nitrogen</u>, <u>phosphorus</u> and <u>potassium</u>) and often <u>small amounts</u> of other elements the plants need, like <u>magnesium</u>.

3) Different kinds of fertilisers have <u>different amounts</u> of these minerals in them — so you can pick the one that's best for your plants.

4) For example, for repairing a lawn with <u>poor root growth</u>, you'd probably want a fertiliser with extra <u>phosphate</u>.

5) If you wanted to grow some <u>big, beautiful rose plants</u>, you may want a fertiliser with more <u>nitrate</u> — as this would increase <u>cell growth</u>.

mmm...nitrogen

Nitrates and phosphates and potassium, oh my...

Mmm... that lovely country smell. It's just animal poo being spread across a field — so, perhaps not that lovely any more. Fertilisers can be great — they help plants grow <u>healthily</u>. But farmers need to be careful when using them — careless use can be <u>harmful</u> to the <u>environment</u> (see page 46).

Intensive Farming

Farmers use what they know about plants to farm more <u>efficiently</u>, and with the world's <u>increasing</u> <u>population</u>, intensive farming could be just what we need — <u>bigger</u> and <u>better</u> yields.

Intensive Farming — Getting the Most Out of Plants and Animals

1) <u>Intensive farming</u> is where farmers try to get <u>as much as possible</u> from their plants and animals.

2) The aim is to produce the maximum amount of <u>food</u> from the <u>smallest possible</u> amount of <u>land</u>, to give a <u>huge variety</u> of <u>quality</u> foods, <u>all year round</u>, at <u>cheap prices</u>. They do this in <u>three</u> main ways...

Intensive Farming Uses Artificial Fertilisers...

1) Plants need <u>certain elements</u>, e.g. <u>nitrogen</u>, <u>potassium</u> and <u>phosphorus</u>, so they can make important compounds like proteins.

2) If plants don't get enough of these elements, their <u>growth</u> and <u>life processes</u> are affected (see previous page).

3) Sometimes these elements are <u>missing</u> from the soil because they've been used up by a <u>previous crop</u>.

4) Farmers use artificial fertilisers to <u>replace</u> these missing elements or provide <u>more</u> of them. This helps to increase the <u>crop yield</u>.

...As Well As Pesticides, Fungicides and Herbicides

1) <u>PESTICIDES</u> are chemicals that kill <u>farm pests</u>, e.g. insects, rats and mice. Pesticides that kill insects are called <u>insecticides</u>. Killing pests that would otherwise eat the crop means there's more left for us.

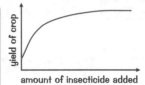

2) <u>FUNGICIDES</u> kill fungi, e.g. moulds that can damage crops.

3) <u>HERBICIDES</u> kill <u>weeds</u>. If you <u>remove</u> plants that compete for the same <u>resources</u> (e.g. nutrients from the soil), it means the crop gets more of them and so grows better.

Animals can be Kept in Controlled Environments

1) In countries like the UK, animals such as <u>pigs</u> and <u>chickens</u> are often <u>intensively farmed</u> (battery farming).

2) They're kept <u>close together indoors</u> in small pens, so that they're <u>warm</u> and <u>can't move about</u>.

3) This saves them <u>wasting energy</u> on movement, and stops them giving out as much energy as <u>heat</u>.

4) This means the animals will <u>grow faster</u> on <u>less food</u>.

5) This makes things <u>cheaper</u> for the farmer, and for us when the animals finally turn up on supermarket shelves.

Intensive farming might just crop up in the exam...

The important stuff is knowing <u>how</u> intensive farming <u>increases</u> the amount of food — <u>fertilisers</u> provide essential <u>minerals</u> for growth. <u>Herbicides</u> remove <u>competition</u>, <u>fungicides</u> prevent disease and putting animals in <u>controlled environments</u> means they <u>waste less energy</u> — all the more for us.

Organic Farming

Intensive farming methods are still used a lot. But people are also using organic methods more and more.

Organic Farming Doesn't Use Artificial Chemicals

An alternative to modern intensive farming is organic farming. Organic methods are more traditional. Where intensive farming uses chemical fertilisers, herbicides and pesticides, organic farming has more natural alternatives.

THE LAND IS KEPT FERTILE BY:

1) Using organic fertilisers (i.e. animal manure and compost). This recycles the nutrients left in plant and animal waste. It doesn't always work as well as artificial fertilisers, but it's better for the environment.

2) Crop rotation — growing a cycle of different crops in a field each year. This stops the pests and diseases of one crop building up, and means nutrients are less likely to run out (as each crop has different needs).

PESTS AND WEEDS ARE CONTROLLED BY:

1) Weeding — physically removing the weeds, rather than just spraying them with a herbicide. Obviously it takes a lot longer, but there are no nasty chemicals involved.

2) Varying crop growing times — farmers can avoid the major pests for a certain crop by planting it later or earlier in the season. This means they won't need pesticides.

3) Using natural pesticides — some pesticides are completely natural, and so long as they're used responsibly they don't mess up the ecosystem.

4) Biological control — Biological control means using a predator, a parasite or a disease to kill the pest, instead of chemicals. For example:

   a) Aphids are pests which eat roses and vegetables. Ladybirds are aphid predators, so people release them into their fields and gardens to keep aphid numbers down.

   b) Certain types of wasps and flies produce larvae which develop on (or in, yuck) a host insect. This eventually kills the host. Lots of insect pests have parasites like this.

   c) Myxomatosis is a disease that kills rabbits. In Australia the rabbit population grew out of control and ruined crops so the myxoma virus was released as a biological control.

Organic Farms Keep Animals in More Natural Conditions

1) For an animal farm to be classified as "organic", it has to follow guidelines on the ethical treatment of animals.

2) This means no battery farming — animals have to be free to roam outdoors for a certain number of hours every day.

3) Animals also have to be fed on organically-grown feed that doesn't contain any artificial chemicals.

Don't get bugged by biological pest control...

The Soil Association is an organisation that certifies farms and products as organic. They have very strict rules about what products can carry their logo — much stricter than the Government's minimum standards. About 70% of all organic food sold in the UK is Soil Association approved.

Comparing Farming Methods

It's all very well knowing how the different farming methods work, but are they actually any good? Both intensive and organic methods have advantages and disadvantages, which I'm afraid you just have to learn.

Intensive Farming is Efficient but can Destroy the Environment

The main advantage of intensive farming methods is that they produce large amounts of food in a very small space. But they can cause a few problems — the main effects are:

1) Removal of hedges to make huge great fields for maximum efficiency.
   This destroys the natural habitat of wild creatures and can lead to serious soil erosion.

2) The same crops are grown year after year (this is called monoculture). This can cause problems, e.g. nutrients required by the crop are removed from the soil — increasing the need for artificial fertilisers. Also pests and diseases build up — this means even more chemicals (pesticides etc.) have to be used.

3) Lots of people think that intensive farming of animals such as battery hens is cruel.

4) The chemicals used can be damaging to the environment:

TOO MUCH FERTILISER — EUTROPHICATION

1) Problems start if some of the fertiliser finds its way into rivers and streams.

2) This happens quite easily if too much fertiliser is applied, especially if it rains afterwards.

3) The result is eutrophication, which can cause serious damage to rivers and lakes.

PESTICIDES DISTURB FOOD CHAINS

1) There's also a danger of pesticides passing through the food chain to other animals.

2) The diagram shows a food chain that ends with otters. The otter ends up with loads of the insecticide because it builds up at each level.

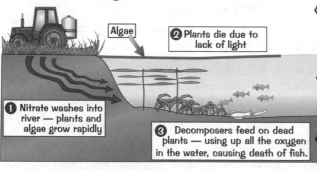

Algae
② Plants die due to lack of light
① Nitrate washes into river — plants and algae grow rapidly
③ Decomposers feed on dead plants — using up all the oxygen in the water, causing death of fish.

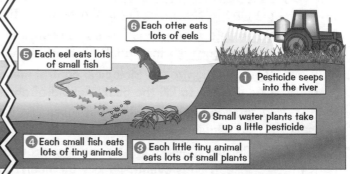

⑥ Each otter eats lots of eels
⑤ Each eel eats lots of small fish
① Pesticide seeps into the river
② Small water plants take up a little pesticide
④ Each small fish eats lots of tiny animals
③ Each little tiny animal eats lots of small plants

Organic Farming Has Advantages and Disadvantages Too

ADVANTAGES

1) Organic farming uses fewer chemicals, so there's less risk of toxic chemicals remaining on food.

2) It's better for the environment. There's less chance of polluting rivers with fertiliser. Organic farmers also avoid using pesticides, so don't disrupt food chains and harm wildlife.

3) For a farm to be classed as organic, it will usually have to follow guidelines on the ethical treatment of animals. This means no battery farming.

DISADVANTAGES

1) Organic farming takes up more space than intensive farming — so more land has to be farmland, rather than being set aside for wildlife or for other uses.

2) It's more labour-intensive. This provides more jobs, but it also makes the food more expensive.

3) You can't grow as much food as you can with intensive farming.

There's nowt wrong wi' spreadin' muck on it...

You need to be able to discuss the advantages and disadvantages of organic and intensive farming. If a question pops up in the exam don't just give your own opinions — its best to give both sides of the argument — even if you do happen to think battery farming is the best thing since sliced bread.

Products from Living Things

As I'm sure you know by now, pretty much all our <u>food</u> comes from <u>living things</u>. It doesn't stop there though, oh no. You'll be amazed at just how much we <u>depend</u> on other living organisms...

Humans Use Living Things For...

1) Food

Humans use plants (e.g. carrots, spinach and wheat) and animals (e.g. cows, chickens and pigs) as <u>sources of food</u>. Plants produce <u>glucose</u> through photosynthesis. Humans can't produce glucose, so the only way we can get it is by <u>eating plants</u> (or by eating animals — which have got their glucose from eating plants or other animals).

2) Clothing

Fabrics like <u>cotton</u> and <u>hemp</u> are made from fibres that come from plants. <u>Leather</u> comes from <u>animal hides</u> (usually cows) and is also used for clothing. The <u>dyes</u> used to <u>colour clothing</u> can also come from plants.

3) Fuels

<u>Wood</u> and other plant material can be <u>burnt</u> as a fuel. <u>Bacteria</u> can be used to produce <u>biogas</u> from plant and animal <u>waste</u>. Biogas can then be burnt to <u>release energy</u>.

4) Medicines

Many of our modern medicines come from living organisms. For example, <u>fungi</u> can make <u>antibiotics</u>. Bacteria can be <u>genetically modified</u> to make a whole range of medicines. <u>Aspirin</u> can be made from a substance extracted from <u>willow bark</u>, and <u>digitalis</u> (a drug for treating heart disease) comes from <u>foxgloves</u>.

5) Construction materials

A whole host of construction materials are derived from living things. Two of the best examples are <u>wood</u> and <u>rubber</u>. Rubber is extracted from <u>rubber trees</u>.

Microbes Can be Used to Make Medicines

1) <u>Antibiotics</u> are chemicals produced by <u>fungi</u> and <u>bacteria</u> that kill other <u>microorganisms</u> (see p.36).

2) The first antibiotic to be discovered was <u>penicillin</u> — it comes from a fungus called <u>Penicillium</u> (and, if you're interested, was discovered by Alexander Fleming in 1929).

3) Scientists discovered the <u>best conditions</u> to grow the mould so that it produces <u>loads</u> of penicillin. It can then be purified and used to treat <u>bacterial infections</u>.

4) Since the discovery of penicillin, lots more <u>antibiotic-producing</u> fungi have been discovered.

5) This is just as well — bacteria <u>evolve rapidly</u>, which means they can become <u>resistant</u> to antibiotics.

6) Another way microbes are used to make medicine is through <u>genetic engineering</u> (see page 49). Bacteria can be modified to produce things like <u>human insulin</u>, which is used to treat <u>diabetes</u>.

Microbes making medicine — now I've heard it all...

It might be hard to believe but those nasty little blighters that can make you <u>ill</u> can also make you <u>better</u> again. Fleming's discovery came along just in time — penicillin was really important in World War II. It was used to treat bacterial infection, which reduced the need for <u>amputations</u> and <u>saved thousands of lives</u>.

Products from Living Things

When you think of <u>food</u> coming from living things I bet bacteria making <u>cheese</u> and fungi making <u>bread</u> aren't the first things that spring to mind.

Bacteria are Used to Make Cheese...

You need to know how cheese is made, so here goes...

1) A culture of <u>bacteria</u> is added to <u>warm milk</u>.
2) The bacteria produce solid <u>curds</u> in the milk.
3) The curds are <u>separated</u> from the liquid whey.
4) <u>More bacteria</u> are added to the curds, and the whole lot is left to <u>ripen</u> for a while.

... And Yoghurt

1) Milk is <u>pasteurised</u> (heated up to 72 °C for 15 seconds) — to kill off any unwanted microorganisms. Then the milk is <u>cooled</u>.
2) A <u>culture</u> of bacteria is added and the mixture is <u>incubated</u> (heated to about 40 °C) in a vessel called a <u>fermenter</u>. The bacteria turn the <u>lactose sugar</u> in the milk into <u>lactic acid</u>. The lactic acid causes the milk to <u>clot</u> and <u>solidify</u> into <u>yoghurt</u>.
3) A <u>sample</u> is taken to make sure it's at the right consistency. Then <u>flavours</u> (e.g. fruit) and <u>colours</u> are sometimes added before the yoghurt is <u>packaged</u>.

YOGHURT

Yeast is Used to Make Bread...

Holes in the bread, which make it nice and light, are made by carbon dioxide bubbles in the dough.

1) Yeast is used in <u>dough</u> to produce nice, light bread.
2) The yeast respires and converts sugars to <u>carbon dioxide</u> and some <u>ethanol</u>. It's the <u>carbon dioxide</u> that makes the bread <u>rise</u>.
3) As the carbon dioxide <u>expands</u>, it gets trapped in the dough, making it lighter.

...And for Brewing Beer and Wine

1) Firstly you need to get the <u>sugar out</u> of the barley or grapes:

BEER

- Beer is made from <u>grain</u> — usually <u>barley</u>.
- <u>Starch</u> in the barley grains is broken down into <u>sugar</u> by <u>enzymes</u>.
- The grain is then <u>mashed up</u> and water is added to produce a <u>sugary solution</u> with lots of bits in it. This is then sieved to remove the bits.
- <u>Hops</u> are added to the mixture to give the beer its <u>flavour</u>.

WINE

The grapes are <u>mashed</u> and water is added... a bit simpler than getting sugar out of barley.

2) <u>Yeast</u> is <u>added</u> and the mixture is <u>incubated</u> (kept warm). The yeast <u>ferments</u> the <u>sugar</u> into <u>alcohol</u>.
3) The beer and wine produced is <u>drawn off</u> through a tap.
4) Sometimes chemicals are added to <u>remove particles</u> and make it <u>clearer</u>.
5) Finally the <u>beer</u> is <u>casked</u> and the <u>wine</u> is <u>bottled</u> ready for sale.

The world's fastest yoghurt — pasteurised before you see it...

Not all microorganisms are bad for you — some of them can be rather <u>helpful</u>. You still need to stop the bad microorganisms <u>infecting</u> your food. That's why milk is <u>pasteurised</u> — <u>heating</u> them kills off microbes, making them <u>perfectly safe</u> and ready for your enjoyment.

Selecting Characteristics

Farmers want their animals and plants to have the best characteristics, e.g. high crop yield or muscular cows for better beef. They can increase the number of 'good' animals or plants they have by one of three methods: 1) selective breeding, 2) genetic engineering or 3) cloning (see next page).

Selective Breeding is Very Simple...

Selective breeding is also called artificial selection, because humans artificially select the plants or animals that are going to breed and flourish, according to what we want from them.

This is the basic process involved in selective breeding:

1) From your existing stock select the ones which have the best characteristics.

2) Breed them with each other.

3) Select the best of the offspring, and combine them with the best that you already have and breed again.

4) Continue this process over several generations to develop the desired traits.

...But Can be Problematic

1) Selective breeding reduces the number of different alleles in a population because the farmer keeps breeding from the same (the "best") animals or plants — they all end up very closely related.

2) This can cause serious problems if a new disease appears — all the plants or animals could be affected.

See page 26 for more on alleles.

3) Because they all have such similar alleles, if one of them doesn't have resistance to a new disease, it's likely none of the others will either — this could easily wipe out a farmer's entire herd or crop.

Genes Can be Transferred into Animals and Plants

1) Scientists can move genes from one organism to another — this is called genetic engineering.

2) 'Foreign' genes (ones from another organism) are transferred into plant or animal cells.

3) The transfer must be done in the very early stages of the plant or animal development (i.e. shortly after fertilisation) so that they develop with the desired characteristics.

4) The characteristics that the plant or animal displays depends on the type of gene inserted.

5) The possibilities are endless, for example: long-life tomatoes can be made by changing the gene that causes fruit to ripen. Animals could be modified to be bigger (increasing meat yield) or to produce new substances, like drugs in their milk.

Genetic Engineering Has Moral and Ethical Issues

All this is nice, but you need to be able to weigh up the benefits against the moral and ethical issues:

1) Some people think it's wrong to genetically engineer other organisms purely for human benefit. This is a particular problem in the genetic engineering of animals, especially if the animal suffers as a result.

2) People worry that we won't stop at engineering plants and animals. Those who can afford it might decide which characteristics they want their children to have, creating a 'genetic underclass'.

3) There are also concerns about 'playing God', and meddling with things that should be left well alone.

4) The long-term evolutionary consequences of genetic engineering are unknown, but there could potentially be quite a few, e.g. seedless fruit — it's seedless so never reproduces sexually. This means no variety so they won't evolve and can't change with the surrounding environment.

Selective breeding — sounds like a night out in my local disco...

Selective breeding is great for improving quality and yield but reduction of the gene pool is a bit of a problem. It's important to remember the advantages and disadvantages and give a balanced answer.

Selecting Characteristics

Eeek, cloning. People get even more worked up about this than they do about genetic engineering. There are two types of cloning used in agriculture — one for animals and one for plants... enjoy.

Cloning is Making an Exact Copy of Another Organism

1) Clones are GENETICALLY IDENTICAL ORGANISMS.

2) Clones occur naturally in both plants and animals. Identical twins are clones of each other. These days clones are very much a part of the high-tech farming industry.

Embryo Transplants in Cows Produce Clones

Normally, farmers only breed from their best cows and bulls. However, such traditional methods would only allow the prize cow to produce one new offspring each year. These days the whole process has been transformed using embryo transplants:

1) Sperm are taken from the prize bull and used to artificially inseminate the prize cow.

2) The fertilised egg divides to give a ball of genetically identical cells, which develops into an embryo.

3) The embryo is taken from the prize cow and split into separate cells. Each cell grows into a new embryo which is a clone of the original one.

4) These embryos are implanted into other cows, called 'surrogate mothers', where they grow.

5) The offspring are clones of each other, NOT clones of their parents.

> You can also produce animals that are clones of their parents. Dolly the sheep was the first mammal to be cloned from an adult animal. The technique is pretty difficult though, so cloned animals aren't produced commercially by this method.

Tissue Cuttings from Plants Produce Clones

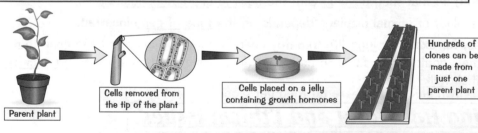

Cells removed from the tip of the plant

Cells placed on a jelly containing growth hormones

Parent plant

Hundreds of clones can be made from just one parent plant

1) First you choose the plant you want to clone based on its characteristics — e.g. a beautiful flower, a good fruit crop.

2) You remove several small bits of tissue from the parent plant.

3) You grow the tissue in a medium containing nutrients.

4) As the tissues produce shoots and roots they can be moved to potting compost to carry on growing.

ADVANTAGE of Cloning	DISADVANTAGE of Cloning
Hundreds of "ideal" offspring can be produced every year from the best bull and cow or the best plants.	The main disadvantage is the same problem that occurs with selective breeding — disease could wipe out an entire population if there are no resistant alleles (see page 49).

Thank goodness they didn't do that with my little brother...

Don't get bogged down in the details of cloning too much — you just need to know why farmers do it and what the disadvantages are. Embryo transplants and tissue cuttings are pretty widely used now — some people are against cloning animals from adult animals, but that's a different kettle of fish altogether.

Revision Summary for Section 2.4

Well, here we are again, the end of another section — yay. By now you should know everything there is to know about agriculture and farming... but there's only one way to find out if you do. Yep, you guessed it — a good old revision summary. If you can't answer the questions then you better go back and learn it again — or else — I'll steal your shoes.

1) Give three differences between plant and animal cells.
2) What is the function of: a) the nucleus, b) the cytoplasm.
3) Which cells are adapted to carry out photosynthesis?
4) How does the structure of root hair cells make them suited to their job?
5) How does being thin and flat make leaves efficient photosynthesisers?
6) What four things do plants need for photosynthesis to happen? Write these in an equation.
7) How can the ideal conditions for photosynthesis be artificially created in glass houses?
8) Give one other advantage of using glasshouses.
9) Why do plants need nitrates?
10) What could happen to a plant if there wasn't enough potassium in the soil?
11) What is magnesium needed for? What might happen to plants without enough magnesium?
12) If you were growing strawberries and wanted to improve the quality of the fruit, which mineral might you add to the soil?
13) What is intensive farming?
14) Why does intensive farming use fertilisers?
15) What is a chemical that's used to kill weeds called?
16) What is battery farming and why is it used?
17) State one alternative to intensive farming.
18) How does this method keep the land fertile?
19) How does the treatment of animals differ between this method and intensive farming?
20) What is meant by 'biological control'?
21) Why can growing the same crops year after year be bad for the environment?
22) How can pesticides disturb food chains?
23) Describe the problems that can be caused in rivers by overuse of fertilisers. For a bonus point, what is the fancy name for this?
24) Give two advantages and two disadvantages of organic farming.
25) Humans can't make their own glucose. Where do they get glucose from?
26) Name two materials used for clothing that are produced from living things.
27) How can bacteria be used to make fuel?
28) What living organisms produce antibiotics?
29) What was the first antibiotic to be discovered?
30) What kind of infections can antibiotics be used to treat?
31) Give an example of how genetic engineering can be used to produce medicines.
32) How can bacteria be used to make cheese?
33) What is pasteurisation and why is it done?
34) When making beer, what does yeast do?
36) Describe the process of selective breeding. Give one disadvantage of it.
37) What is genetic engineering? Describe the ethical issues surrounding it.
38) Describe the advantages and disadvantages of using cloning in agriculture.

Useful Chemicals from the Ground

Most of the <u>substances we use</u> can be found in the <u>Earth</u> — on their <u>own</u> or <u>in rocks</u>. Elements like sulfur and gold can occur naturally on their own, as can compounds like limestone. But some useful substances are in mixtures or compounds in rocks and minerals, e.g. <u>salt</u> in <u>rock salt</u> and <u>metals</u> in <u>metal ores</u>. But anyway, as to what <u>elements</u>, <u>compounds</u> and <u>mixtures</u> are... well, if you don't know, read this page.

Elements are Made Up of Just One Type of Atom

1) As you probably know, <u>all</u> <u>substances</u> are made up of <u>atoms</u> (if you didn't know then see Section 2.6 — it's all about atoms).

2) If a substance only contains <u>one type</u> of atom it's called an <u>element</u>. Quite a lot of everyday substances (like copper and gold) are <u>elements</u>.

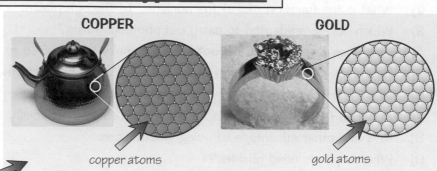

COPPER

GOLD

copper atoms

gold atoms

Compounds and Mixtures Contain Different Types of Atom

1) If there's <u>more than one type</u> of atom, e.g. copper and oxygen, it's either a <u>mixture</u> or a <u>compound</u>.

2) A <u>compound</u> is where different atoms have <u>bonded together</u> chemically, e.g. salt and limestone.

3) A <u>mixture</u> contains different substances that are <u>not</u> chemically bonded together, e.g. rock salt.

4) <u>Mixtures</u> are generally <u>easier to separate</u> than compounds.

<u>SALT</u> is a <u>COMPOUND</u> of sodium and chlorine, called sodium chloride.

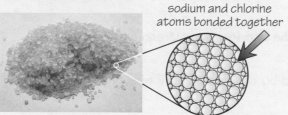

sodium and chlorine atoms bonded together

<u>LIMESTONE</u> is a <u>COMPOUND</u> of calcium, carbon and oxygen, called calcium carbonate.

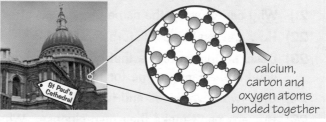

calcium, carbon and oxygen atoms bonded together

<u>ROCK SALT</u> is a <u>MIXTURE</u> of two compounds — salt and sand.

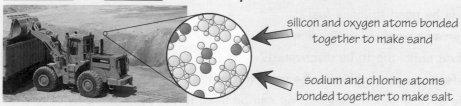

silicon and oxygen atoms bonded together to make sand

sodium and chlorine atoms bonded together to make salt

See p.54 for how to separate out salt.

So next time you're at St Paul's Cathedral eating salty chips...

...you can thank the Earth for making it all possible (or not — up to you). Cover the rest of the page and fill in this table with the substances below: salt, rock salt, oxygen, crude oil, limestone, gold.

Elements	Compounds	Mixtures

Useful Chemicals from the Ground

We can <u>mine</u> or <u>quarry</u> to take useful substances out of the ground. Some of these materials can be used <u>as they are</u>, but many are just the <u>starting material</u> to make other useful substances — by <u>separating</u> them into their constituent parts (see p.54-55) or by <u>combining</u> them with other materials.

Sulfur and Gold Can be Found in the Ground As They Are...

1) <u>Sulfur</u> is the <u>yellow</u>, powdery stuff you find in the ground near <u>volcanoes</u> and in <u>geothermal areas</u> that gives off a nasty <u>rotten-egg-smelling gas</u>. It's a non-metal <u>element</u> (see p.71).

2) Sulfur's <u>not very reactive</u>, so it often occurs naturally <u>on its own</u> in the Earth, rather than reacting with other substances.

3) Among the many <u>uses</u> of sulfur are for making <u>fertilisers</u>, <u>sulfuric acid</u> for <u>petrol refining</u> and <u>car batteries</u>, <u>gunpowder</u> for guns (no really) and for <u>fireworks</u>.

A smoking, sulfur-rimmed crater in New Zealand. It smells.

1) <u>Gold</u> is a <u>metallic element</u>. Like sulfur, gold is very <u>unreactive</u>, so it's usually found <u>on its own</u> in the ground, in <u>shiny gold bits</u>.

2) It can be exposed by things like river erosion, so you can sometimes find it in river beds. People used to '<u>pan</u>' for gold — they just <u>sifted</u> through the <u>mud and silt</u> from the river bed, and picked out the <u>shiny grains</u>. Then they melted it all down to make nice tiaras and things.

3) The main <u>uses</u> of gold are for <u>jewellery</u> (clearly), and in situations where you want a metal that <u>won't react</u> with anything, such as <u>tooth fillings</u> and in <u>electric circuits</u>.

...And So Can Limestone and Marble

<u>Limestone</u> and <u>marble</u> are both natural forms of the compound <u>calcium carbonate</u> that can be <u>mined</u> straight from the ground.

LIMESTONE

Limestone is a rough, pale grey rock that's <u>easy to shape</u> into blocks for <u>building</u> with. It's also used to make <u>cement</u>, <u>glass</u> and <u>lime</u> (to put on acid soil). For more on the uses of limestone see page 73-74.

Uses of Limestone

MARBLE

Marble is a <u>bright white</u>, crystalline rock, often with veins of other minerals. It can be <u>polished</u> to a <u>shiny</u>, smooth finish, and can make impressive <u>statues</u>, <u>sculptures</u>, <u>palaces</u>, etc.

Uses of Marble

I feel like a stroll round a volcano — to get out in the elements...

Some substances are useful in the form you find them. You just dig them out, give 'em a bit of a spit and polish, and Bob's your uncle: St Paul's Cathedral. Well, almost.

54

Rock Salt and Crude Oil

Unfortunately not everything we dig up from the ground puts up as little _resistance_ as _gold_ and _marble_. Oh no, the likes of _salt_ and _oil_ take a little more coaxing...

Rock Salt Needs to be Refined...

1) Rock salt is a _mixture_ of _salt_ and _sand_. It can be used for _de-icing roads_ — the salt _melts_ the ice and the sand provides _grip_.

2) The _salt_ in rock salt has more _uses_ — so the sand is usually _filtered out_ of the salt-sand mixture to leave _refined salt_.

3) Refined salt (also known as _table salt_ — NaCl) is used to flavour food.

4) In Britain there are large _underground deposits_ of rock salt, left from _ancient seas_. (In _hot countries_ they get salt in a different way — they pour _sea water_ into large open containers and the water _evaporates_ to leave _salt_.)

... As Does Crude Oil

1) Crude oil provides _fuel_ for most modern transport.

2) It also provides the _raw materials_ for making various _chemicals_, including _plastics_.

3) Crude oil is _extracted_ from the ground by _drilling_. Underground oil fields are found all over the place, for example in the Middle East. Many are also found offshore, which means the oil has to be drilled up from the _sea floor_.

4) The oil is formed from the buried remains of _plants and animals_ — so it's a _fossil fuel_ (see p.56). Over millions of years, with _high temperature_ and _pressure_, the remains turn to crude oil.

5) Crude oil is a _mixture_ of _lots of chemicals_, but different modes of transport use different combinations of these chemicals. This is why it needs to be _refined_ — to _separate_ it into _different chemicals_.

6) It's refined using _fractional distillation_:

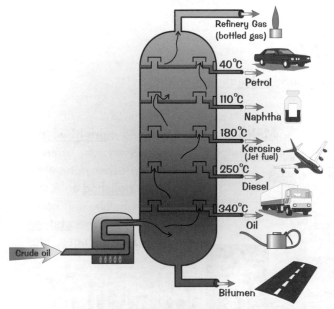

1) The crude oil is _heated up_ and the different chemicals _evaporate_ and rise up the column.

2) The different chemicals _condense_ at different _temperatures_ and can be collected.

3) The process works _continuously_, with heated crude oil piped in at the _bottom_ and the various _chemicals_ being _constantly collected_ at the different levels.

7) It's important to remember that oil can be seriously bad news for the _environment_ — oil slicks at sea, old engine oil down the drain, plastics that won't rot if you throw them away. See page 63 for more on the _environmental impacts_ of fossil fuels.

Someone threw some NaCl at me — I said, "hey, that's a salt"...

Back in the olden days when they didn't have fridges, salt was used to _preserve food_. People also used to be paid in salt — the word salary comes from the Latin word for salt (sal). Bet you didn't know that.

Section 2.5 — Mining & Pollution   Countryside & Environmental Management

Metals

It's not often you find big lumps of metal in the ground — the metal atoms tend to be joined to other atoms in <u>compounds</u> (see p.52) Separating the metal out can be a <u>tricky</u> and <u>expensive</u> process...

Ores Contain Enough Metal to Make Extraction Worthwhile

1) A <u>metal ore</u> is a naturally occurring rock which contains <u>enough metal</u> to make extraction from the ore <u>worthwhile</u>.

2) In many cases the ore is an <u>oxide</u> of the metal (it's bonded to oxygen). But not all ores are oxides. Here are a few examples:

- One type of <u>iron ore</u> is called <u>haematite</u>. This is iron oxide (Fe_2O_3).
- The main <u>aluminium ore</u> is called <u>bauxite</u>. This is aluminium oxide (Al_2O_3).
- Another type of <u>iron ore</u> is <u>pyrite</u>. This is iron disulfide (FeS_2).

Pyrite — also known as fool's gold.

3) Ores are "finite resources" — there's a <u>limited amount</u> of them. Once you've dug them all up there's no more.

Some Metals Can be Extracted by Reduction with Carbon

Some metals can be <u>extracted</u> by <u>heating</u> with <u>carbon</u> or <u>carbon monoxide</u> (the fancy name for this is reduction). <u>Reduction</u> simply means that the <u>oxygen</u> is <u>removed</u> — this makes it a 'pure' metal.

1) <u>Carbon</u> and <u>carbon monoxide</u> are <u>reducing agents</u> (they steal the oxygen away from the metal).

2) This only works for metals that are <u>less reactive</u> than <u>carbon</u> though.

3) Two good examples are the reduction of <u>iron oxide</u> to <u>iron</u>, and <u>lead oxide</u> to <u>lead</u>. The end product for both reactions is the <u>metal</u> and <u>carbon dioxide</u>.

iron oxide + carbon monoxide → iron + carbon dioxide
$$Fe_2O_3 + 3CO \rightarrow 2Fe + 3CO_2$$

lead oxide + carbon → lead + carbon dioxide
$$PbO_2 + C \rightarrow Pb + CO_2$$

These are symbol equations — see p.67 for more.

Mining Can Have an Effect on the Environment

1) Mining metal ores can be <u>good</u> — many <u>useful products</u> can be made. It also provides local people with <u>jobs</u> and brings <u>money</u> into the area. This means services such as <u>transport</u> and <u>health</u> can be improved.

2) <u>But</u> mining ores is <u>bad for the environment</u> — it causes noise, scarring of the landscape and loss of habitats. Abandoned mine shafts can also be <u>dangerous</u> because they are so deep.

You won't catch me looking at gold in awe...

Remember, when these ores are <u>reduced</u> the <u>oxygen</u> is removed. If a metal is <u>more reactive</u> than carbon then it can't be extracted by reduction. They could ask you a question on reducing these ores in the exam and might expect you to write an <u>equation</u> — just remember the end product is always the <u>metal</u> and <u>carbon dioxide</u>.

56

Fossil Fuels

Fossil fuels are awesome — they provide a lot of the energy needed to make electricity. I don't know what we'd do without them — the problem is, one day we may have to do without them. It won't be long before we all have giant hamster wheels in our houses to make our own electricity just so we can watch Corrie.

The Energy in Fossil Fuels Comes from the Sun

Fossil fuels include fuels such as coal, oil and natural gas. Fossil fuels formed over millions of years from the remains of dead plants and animals (which originally got all their energy from the Sun). Fossil fuels are a useful source of energy (see p.61 for how they're used to generate electricity). Energy is released from fossil fuels by burning them with oxygen.

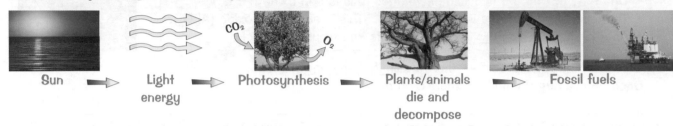

Sun → Light energy → Photosynthesis → Plants/animals die and decompose → Fossil fuels

Combustion of Fossil Fuels Releases Energy

1) Fossil fuels are hydrocarbons — this means that they're made up of hydrogen and carbon atoms.

2) The complete combustion of any hydrocarbon in oxygen will produce only carbon dioxide and water as waste products:

$$\text{hydrocarbon} + \text{oxygen} \rightarrow \text{carbon dioxide} + \text{water} (+ \text{energy})$$

Fossil Fuels Can Cause Environmental Problems

Even though fossil fuels provide us with lots of energy, there are quite a few problems associated with burning them...

1) Fossil fuels release CO_2 — this adds to the greenhouse effect and contributes to climate change. This is bad news as increased temperatures lead to melting of the polar ice caps and rising sea levels. (There's more on climate change and the greenhouse effect on page 63.)

2) Fossil fuels often contain impurities such as sulfur. Burning coal and oil releases the sulfur as sulfur dioxide, which causes acid rain. This can be reduced by taking the sulfur out before it's burned, or by cleaning up the emissions.

3) Coal mining makes a mess of the landscape (see p.55 for more on mining).

4) Oil spillages cause serious environmental problems. We try to avoid them, but they always happen.

5) Another problem is that there are limited deposits of fossil fuels in the Earth's crust. Eventually they will run out and because they take millions of years to form, we'll be waiting quite a while before there are any more.

Revising fossil fuels — you know the drill...

Just think, the next time you put the kettle on, the energy could well have come from some prehistoric animal, that ate a plant, that converted light energy from the Sun, and now you're using it to make a brew — pretty amazing I think. You may not have come across the word hydrocarbon before, but all it means is a substance containing only hydrogen and carbon atoms — pretty obvious when you think about it really. If you don't know much about atoms, don't fear — the next section is all about them.

Alternatives to Fossil Fuels

Alternative energy isn't just for hippies. Many see renewable resources as the only way to stop climate change, however they do also have a number of disadvantages...

Renewable Energy Resources Will Never Run Out

The renewables are: 1) Wave   2) Tidal   3) Wind   4) Hydroelectric   5) Solar

- These will never run out.
- Most of them do damage the environment, but in less nasty ways than non-renewables.
- But they don't all provide much energy and some are unreliable because they depend on the weather.

Wave Power — Lots of Little Wave Converters

1) Wave generators can be located around the coast.
2) As waves come in to the shore they provide an up and down motion — this movement can be used to drive a generator.
3) There's no pollution. The main problems are spoiling the view and being a hazard to boats.
4) They're fairly unreliable though, since waves tend to die out when the wind drops.
5) Initial costs are high but there are no fuel costs and minimal running costs.

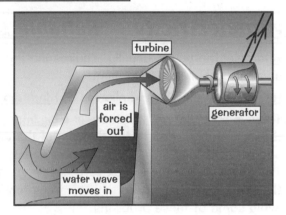

Tidal Barrages — Using the Sun and Moon's Gravity

1) Tidal barrages are big dams built across river estuaries, with turbines in them.
2) As the tide comes in it fills up the estuary to a height of several metres. This water can then be allowed out through turbines at a controlled speed. It also drives the turbines on the way in.
3) There's no pollution. The main problems are preventing free access by boats, spoiling the view and altering the habitat of the wildlife, e.g. wading birds, sea creatures and beasties who live in the sand.
4) Tides are pretty reliable in the sense that they happen twice a day without fail. The only drawback is that the height of the tide is variable — so lower tides will provide less energy.
5) Initial costs are moderately high but there are no fuel costs and minimal running costs. Even though it can only be used in a few of the most suitable estuaries tidal power has the potential for generating a significant amount of energy.

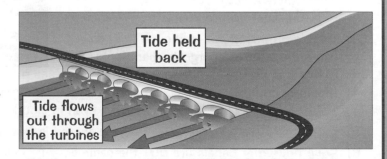

Renewable energy — wave goodbye to fossil fuels...

It's really important that you appreciate the big big differences between tidal power and wave power. Yes, they both involve salty sea water, but the similarities start and end there. Make sure you remember this for the exam — you could be throwing away some easy marks if you get the two mixed up.

Alternatives to Fossil Fuels

There you have it, turns out waves not only come in handy for surfing but also for generating electricity — who'd have guessed. They aren't the only sources of renewable energy though — here's another three for your enjoyment...

Wind Power — Lots of Little Wind Turbines

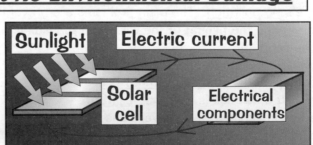

1) This involves putting lots of windmills (wind turbines) up in exposed places like on moors or round coasts.

2) Wind turns the blades, which turn a generator inside the turbine.

3) There's no pollution (except for a little bit when they're made).

4) There are however a few problems. You need about 5000 wind turbines to replace one coal-fired power station. They spoil the view and can be noisy, which can be annoying for people living nearby. There's no power when the wind stops, and it's impossible to increase supply when there's extra demand.

Hydroelectricity — Reliable but Damages the Environment

1) Hydroelectric power usually requires the flooding of a valley by building a big dam.

2) Rainwater is caught and allowed out through turbines.

3) There is no pollution, but there is a big impact on the environment due to the flooding of the valley and possible loss of habitat for some species. The reservoirs can also look very unsightly when they dry up. Location in remote valleys tends to avoid some of these problems.

4) A big advantage is immediate response to increased demand, and there's no problem with reliability except in times of drought.

5) Initial costs are high, but there's no fuel and minimal running costs.

Solar Energy — Expensive but No Environmental Damage

1) Solar cells generate electric currents directly from sunlight.

2) They're often used in remote places where there's not much choice, e.g. the Australian outback.

3) There's no pollution. (Although they do use quite a lot of energy to manufacture in the first place.)

4) In sunny countries solar power is a very reliable source of energy — but only in the daytime. Solar power can still be cost-effective even in cloudy countries like Britain.

5) Initial costs are high but after that the energy is free and running costs almost nil.

6) Solar hot water panels are NOT the same as solar cells. Solar hot water panels are used to directly heat water for household use — there's no electricity involved.

The hydroelectric power you're supplying — it's electrifying...

The big advantage of renewable fuels is that they don't release CO_2. Like all good things in life, they also have a number of disadvantages, such as unreliability and environmental damage. They could ask you to compare fossil fuels and a renewable source — make sure you know the pros and cons.

Alternatives to Fossil Fuels

In addition to <u>renewable sources</u> there's also a <u>non-renewable</u> alternative to fossil fuels — <u>nuclear</u>. It's <u>clean</u>, <u>readily available</u> and the fuel is <u>cheap</u>. Despite all this, there are those who <u>disagree</u> with it because of its potential impact on the environment.

Nuclear Energy Comes from the Nuclei of Atoms

1) <u>Nuclear power stations</u> are powered by <u>nuclear reactors</u>.
2) In a nuclear reactor the 'fuel' is <u>uranium</u> or <u>plutonium</u>. The 'fuel' isn't burnt in air as fossil fuels are. Instead the 'fuel' atoms are <u>split up</u> (this is known as <u>nuclear fission</u>) — this <u>releases heat energy</u>.
3) This energy is used to <u>heat water</u>, to drive a <u>steam turbine</u>, to generate electricity (see p.61).
4) <u>1 kg of uranium</u> can give out as much heat as <u>2 000 000 kg of coal</u>.

Nuclear Power Stations Don't Emit CO_2...

1) One of the biggest problems with <u>fossil fuels</u> is that they give out <u>carbon dioxide</u> as they burn (p.56).
2) <u>Carbon dioxide</u> adds to the <u>greenhouse effect</u>, causing <u>global warming</u>.
3) Nuclear power produces <u>no CO_2</u>, though there are other <u>environmental problems</u> (see below).
4) It's <u>non-renewable</u> but there's enough fuel to last <u>hundreds</u> of years, making it a more <u>long-term solution</u> than fossil fuels.

...but Radioactive Waste is an Environmental Risk

<u>Nuclear power</u> is <u>clean</u> but the <u>nuclear waste</u> is very <u>dangerous</u> and difficult to <u>dispose of</u>.

1) Nuclear power produces <u>radioactive waste</u> — this can emit dangerous radiation for <u>thousands of years</u>, which is why nuclear waste needs to be <u>stored safely</u>.
2) Storage depends on the <u>type of waste</u> — some types of waste are <u>more dangerous</u> than others, e.g. <u>gloves</u> or <u>clothing</u> used in a lab are fairly safe, whereas <u>spent fuel rods</u> have to be <u>cooled in water</u> before being <u>sealed in glass blocks</u> and <u>buried underground</u>.

When they're working <u>normally</u> nuclear reactors are very <u>safe and clean</u>, but nuclear power always carries the <u>risk</u> of a <u>major catastrophe</u> like the <u>Chernobyl disaster</u>.

1) In 1986 technicians at a reactor in Chernobyl turned off the <u>safety devices</u> to test the reactor.
2) The reactor <u>overheated</u> and <u>exploded</u>. Large amounts of <u>radioactive material</u> were released into the <u>atmosphere</u> and spread across Europe.
3) Many people <u>died</u> and many more were made <u>ill</u>. Many areas around Chernobyl are still <u>contaminated</u> to this day.

Nuclear <u>fuel</u> (i.e. uranium) is <u>relatively cheap</u>, but the <u>overall cost</u> of nuclear power is <u>high</u> (because of the cost of the <u>power plant</u> and <u>disposal</u> of waste). Plants also need to be well <u>maintained</u> — <u>accidents</u> are more likely to happen as the reactors become <u>older</u> and <u>less efficient</u>.

Nuclear — think that's the name of my face wash...

With <u>decreasing</u> supplies of <u>fossil fuels</u> and <u>renewables</u> unable to meet all our <u>energy needs</u>, <u>nuclear</u> could be the only viable option for the <u>future</u>. That doesn't mean everyone has to like it though — many people <u>oppose</u> nuclear power because of the <u>environmental damage</u> and the possibility of an explosion.

Comparing Energy Resources

Just in case you haven't quite had enough on the different types of energy resources, here's another page on it. There shouldn't be too much that comes as a shock to you on this page — it's more like a bit of a summary of the last four pages to help you put it all together and see the bigger picture.

Environmental Issues

If there's a fuel involved, there'll be waste pollution and you'll be using up resources.

If it relies on the weather, it's often got to be in an exposed place where it sticks out like a sore thumb.

Visual Pollution
Coal, Oil, Gas, Nuclear, Tidal, Waves, Wind, Hydroelectric

Atmospheric Pollution
e.g. Coal, Oil, Gas

Other Problems
Nuclear (dangerous waste, explosions, contamination), Hydroelectric (dams bursting)

Using Up Resources
Coal, Oil, Gas, Nuclear

Noise Pollution
Coal, Oil, Gas, Nuclear, Wind

Disruption of Leisure Activities (e.g. boating)
Waves, Tidal

Disruption of Habitats
Hydroelectric, Tidal

Reliability Issues

All the non-renewables are reliable energy providers (until they run out).

Many of the renewable sources depend on the weather, which means they're pretty unreliable here in the UK. The exception is tidal power.

Running/Fuel Costs

Renewables usually have the lowest running costs, because there's no actual fuel involved.

Set-Up Costs

Renewable resources often need bigger power stations than non-renewables (for the same output) — the bigger the power station, the more expensive.

Nuclear reactors and hydroelectric dams also need huge amounts of engineering to make them safe, which bumps up the cost.

Location Issues

Power stations have to be near the stuff they run on.

Solar — pretty much anywhere, though the sunnier the better.

Gas — pretty much anywhere there's piped gas.

Hydroelectric — hilly, rainy places with floodable valleys, e.g. the Lake District, Scottish Highlands.

Wind — windy places like moors, coasts or out at sea.

Oil — near the coast (oil is transported by sea).

Waves — on the coast.

Coal — were built near coal mines (new ones are near ports as we import most of our coal now).

Nuclear — away from people (in case of disaster), near water (for cooling).

Tidal — big river estuaries where a dam can be built.

Set-Up Time

This is affected by the size of the power station, the complexity of the engineering and also the planning issues (e.g. discussions over whether they should be allowed to build a nuclear power station on a stretch of beautiful coastline can last years).

Gas is one of the quickest to set up.

The biggest problem is we need too much electricity...

It would be lovely if we could get rid of all the polluting power stations and replace them all with clean, green power sources, just like that... but it's not quite that simple. Renewable energy has its problems too, and probably couldn't power the whole country without having a wind farm in everyone's backyard.

Generating Electricity

As you've already seen, there are loads of different types of <u>fuels</u> out there and they're a great source of <u>energy</u> for us. But just how do we <u>use</u> that energy? <u>Fire</u> may have satisfied the needs of prehistoric people (mostly because they just liked looking at bright things), but let's face it they didn't have way cool computers, plasma tellies or hairdriers. We need to <u>convert</u> the energy into a <u>useful</u> form — electricity...

Most Power Stations Use Steam to Drive a Turbine

1) <u>Most</u> of the electricity we use is <u>generated</u> from <u>non-renewable</u> sources of energy (<u>coal</u>, <u>oil</u>, <u>gas</u> and <u>nuclear</u>).

2) <u>Big power stations</u> all work on the same principles — using <u>steam</u> to <u>turn</u> a <u>turbine</u>.

3) The big <u>difference</u> between power stations is the <u>boiler</u>. The boiler in a <u>coal-powered</u> power station will be very different from the reactor in a <u>nuclear</u> power station.

4) There are <u>three basic stages</u> to generating electricity in a typical power station:

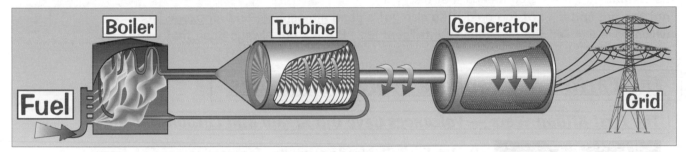

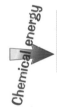 Chemical energy

1) Fuel is burnt and the <u>heat given out</u> is used to <u>boil water</u>.

Heat energy

2) <u>Steam</u> from the boiling water is used to <u>turn a turbine</u>.

Kinetic energy

3) The <u>turbine rotates</u> a <u>generator</u> to <u>create electricity</u>.

Electrical energy

5) Before becoming <u>electrical energy</u> the <u>chemical energy</u> is converted into two other types of energy (<u>heat energy</u> and <u>kinetic energy</u>).

6) This process <u>isn't very efficient</u> because every time the energy is <u>converted</u> you <u>lose a bit</u>.

7) This means that electricity can be quite <u>expensive</u> to generate (and so expensive to buy).

Electricity Gets Around via the National Grid

The <u>National Grid</u> is the <u>network</u> of pylons and cables that covers <u>the whole of Britain</u>, getting electricity to homes everywhere. Whoever you pay for your electricity, it's the National Grid that gets it to you.

1) The <u>National Grid</u> takes electrical energy from the <u>power stations</u> to everywhere it's needed, like <u>homes</u> and <u>industry</u>.

2) It means power can be <u>generated</u> anywhere on the grid, and then be <u>supplied</u> anywhere else on the grid.

S-team drives a turbine, the A-team drove a van...

The way we make electricity today uses the same idea as <u>steam engines</u> — heating <u>water</u> to make <u>steam</u> and then using steam to produce <u>kinetic energy</u>. Steam engines were invented way before you and I were born, in the 17th century. It's amazing that the same <u>principle</u> is still being used today — bet you can't think of many other things that have lasted that long (apart from maybe forks and Bruce Forsyth).

The Origins of Life on Earth

Before the evolution of plants the atmosphere was mostly filled with carbon dioxide. Thankfully it's not like that any more, the present atmosphere contains only 0.04% carbon dioxide — far more appealing. One of the most important factors in the evolution of life is the Earth's position in the Solar System.

The Earth is Habitable Because of Where It is

Our Solar System consists of a star (the Sun) and lots of stuff orbiting it.

- Closest to the Sun are the inner planets — Mercury, Venus, Earth and Mars.
- Then come the outer planets — Jupiter, Saturn, Uranus and Neptune.

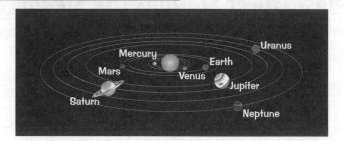

The Earth is an ideal location for life — its distance from the Sun means that it's not too hot and it's not too cold. It's also got a good atmosphere and an ozone layer, which have enabled the evolution of intelligent life. But it's not always been that way...

The Earth's Atmosphere Evolved in Stages:

The First Billion Years — Volcanoes Gave Out Steam and Carbon Dioxide

1) The Earth's surface was molten for many millions of years. Any atmosphere just boiled away.
2) Eventually it cooled and a thin crust formed, but volcanoes kept erupting — releasing mainly carbon dioxide, but also some steam.
3) The early atmosphere was mostly CO_2 (virtually no oxygen).

Volcanoes played an important part in the formation of the Earth's surface and its atmosphere. Volcanoes erupt because of the movement of tectonic plates (large flat chunks of rock that make up the Earth's surface) — which also causes earthquakes. Environmental scientists constantly monitor the movement of tectonic plates to try and predict when an earthquake will occur or a volcano will erupt.

The Next Two Billion Years — Green Plants Evolved and Produced Oxygen

1) Green plants evolved over most of the Earth.
2) The green plants steadily removed CO_2 and produced O_2 by photosynthesis.
3) Much of the CO_2 from the air became locked up in fossil fuels and rocks.

Over the last two billion years or so leading up to the present day oxygen levels increased, allowing the evolution of intelligent life. Oxygen also created the ozone layer, which blocks the Sun's harmful rays. Nowadays there's hardly any carbon dioxide left in the atmosphere.

Volcanoes are just angry mountains — they're fuming...

And they said it couldn't be done — the history of life on Earth on one page, and to think it took some guy called Charles Darwin a whole book just to write about the origin of species. The Solar System and the evolution of the atmosphere seem like funny topics to go on the same page but I hope you can see why they're both here — if the Earth wasn't where it is then life wouldn't have been able to evolve at all.

Human Impact on the Earth

The world's population is increasing — and as it does industrial activity must also increase to keep up with the demand for goods and energy. The problem is, with the increase in industrial activity also comes an increased impact on the Earth and its atmosphere.

Lots of People Need Lots of Energy

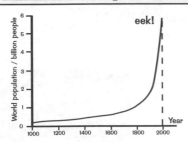

1) The population of the world is currently around six and a half billion. It has increased rapidly over the last few hundred years and it shows no sign of slowing.

2) As the population increases our effect on the planet also increases — we're extracting more and more from the Earth and producing more energy just to meet the demands of an ever-increasing population.

Burning More Fossil Fuels Increases Climate Change...

1) The more energy we need the more fossil fuels we have to burn (at least until there are more alternatives). This produces more CO_2 — a greenhouse gas.

Cars also produce CO_2

2) Greenhouse gases in the atmosphere act like an insulating layer, but if there's too much they'll cause the Earth to heat up.

3) The Earth is gradually heating up because of increasing levels of greenhouse gases — this is causing the climate to change.

4) Climate change has scientists quite worried — the consequences could be serious. Increased temperatures are causing sea levels to rise — bad news for people in low-lying areas. There could also be problems with changing weather patterns leading to droughts, flooding etc.

5) It's not only human activity that releases CO_2 into the atmosphere though. The high concentration of CO_2 in the early atmosphere was caused by volcanoes (see previous page) — they're still responsible for releasing large amounts of CO_2 into the atmosphere today.

The climate is such a complicated system that it's not possible to predict exactly what will happen as temperatures rise. Scientists monitor changes in temperature and CO_2 levels so that better predictions can be made.

...And Acid Rain

1) Sulfur dioxide and nitrogen oxides cause acid rain.

2) They come from car engines and burning fossil fuels.

3) When these gases mix with rain clouds they form dilute acids.

4) So the more fossil fuels we burn for energy and the more cars we have, the more acid rain is caused.

5) Acid rain can cause a lake to become more acidic. Many organisms are sensitive to changes in pH and can't survive in more acidic conditions. Many plants and animals die.

6) Acid rain can also kill trees and damage limestone buildings and statues.

Environmental scientists monitor acid rain by measuring the pH of water in seas, lakes and rivers.

I always wanted to grow bananas in my backyard...

There are so many ifs and buts with climate change. Britain could get hotter or colder, wetter or drier — the truth is, nobody really knows exactly what's going to happen. Scared? You should be.

Human Impact on the Earth

A bigger population also needs more stuff, which means we need to dig up more from the ground.
We're coping OK at the moment but just wait until every person in China has a plasma TV and drives a car.

Living Better Uses More Resources...

As well as increasing in size, the population is also increasing in wealth. Wealthier people are buying more and more products — to produce these we need more resources and more manufacturing.

...Which Means More Mining...

In many countries mining is now regulated by strict guidelines, but it still has the potential to damage the environment through things like contamination of land and water, destruction of habitats, loss of wildlife and noise pollution.

...And Industrial Activity...

Once the raw materials have been extracted from the ground, they have to be turned into desirable products and then transported to where there is a market. But increased industrial activity produces more industrial waste — some of this can be damaging to the environment.

1) Industrial chemicals — these can run into rivers and streams, where they're taken up by organisms at the bottom of the food chain. Many of these chemicals aren't broken down by the organisms, so when they're eaten the chemical is passed on. The concentration of the chemical increases as it passes up the food chain, which can kill the animal or fish at the top (see p.46).

2) Oil — spills from oil tanker accidents and also oil from boat engines harm water life.

3) Some metals (e.g. lead and mercury) are poisonous. They can get into the water supply from old lead pipes or careless waste disposal.

Scientists monitor water pollution by measuring things like the pH of water, the nutrient content, the amount of gases in the water and by looking at the plant and animal life.

...And Waste

As we produce more and more things we also produce more and more waste. You've already seen how the release of waste gases and chemicals can affect air and water, but it can also have an effect on land.

1) In England and Wales alone household, industrial and commercial waste adds up to 100 million tonnes per year, and is increasing by 3% each year.

2) The UK currently recycles around a quarter of household waste. Waste that isn't recycled mostly goes to landfill sites — but some waste is toxic, which means the land becomes polluted.

3) The more waste that can be recycled the better — it means that less land is polluted and fewer materials have to be manufactured or extracted to make new products, saving resources and energy.

Water pollution. Do your bit — stop weeing in the sea...

It's all quite serious really, and it could get even worse — it won't be long before the world is flooded and we all develop gills, float around on home-made boats drinking our own wee — oh no, sorry, that was Water World. But it's still looking a bit grim, unless we get our act together who knows what will happen.

Revision Summary for Section 2.5

Just when you were getting into the swing of stuff that's dug up from the ground we went and caught you off guard by throwing in the origin of life on Earth and how us humans have mucked up the planet. Just to make sure you've actually learnt all this stuff (and not been too busy looking at all the pretty pictures), why not have a go at these questions? Go on — you know you want to.

1) How many types of atoms does an element contain?
2) Give two examples of elements.
3) What is meant by the terms compound and mixture? Give an example of each.
4) Why does gold occur uncombined in the ground?
5) Limestone and marble are natural forms of which compound?
6) Give one use of limestone and one use of marble.
7) In Britain where do we get rock salt from?
8) How is table salt refined from rock salt?
9) Give two uses of crude oil.
10) What is crude oil made from?
11) Describe the process of fractional distillation.
12) What is a metal ore?
13) What is reduction? Give two examples of reducing agents.
14) How could you extract iron from iron oxide? Write a symbol equation for the reaction.
15) Give one advantage and one disadvantage of mining.
16) Where does the energy stored in fossil fuels originally come from?
17) What are hydrocarbons? Write a word equation for the complete combustion of a hydrocarbon.
18) Give three environmental problems associated with fossil fuels.
19) Give one general advantage and one general disadvantage of renewable energy resources.
20) Give one advantage and one disadvantage of: a) tidal power, b) wave power, c) wind power, d) hydroelectric power, e) solar.
21) What type of energy is produced from nuclear fission?
22) Why must nuclear waste be disposed of properly?
23) Why is the overall cost of nuclear power quite high?
24) What factors influence the choice of energy resource for a power station?
25) Describe how a typical power station generates electricity.
26) Why is electricity expensive to generate?
27) How does electricity get from a power station to where it's needed?
28) Where in the Solar System is the Earth located? How has this location enabled the evolution of life?
29) What caused high levels of CO_2 in the atmosphere in the first billion years?
30) What increased the level of O_2 in the atmosphere?
31) How are humans increasing CO_2 levels in the atmosphere?
32) How does the increase in population affect energy production?
33) What effect does increased CO_2 in the atmosphere have?
34) Where do the gases that cause acid rain come from?
35) How might increased industrial activity cause environmental damage?

Chemical Building Blocks

You might not realise it, but building a modern home involves a fair bit of science.
For example, <u>building materials</u> need to have <u>different properties</u> depending on what they're used for.
And a material's <u>properties</u> depend on the <u>atoms</u> it contains, and the way those atoms are <u>held together</u>.

Atoms Have a Small Nucleus Surrounded by Electrons

Pick an object... <u>any</u> object. Whatever you've chosen, it's made up of <u>atoms</u> (since <u>everything</u>'s made up of atoms). And if you could look close enough, you'd see that all those atoms look a bit like the big pic below — with even tinier particles called <u>protons</u>, <u>neutrons</u> and <u>electrons</u> arranged in a particular way.

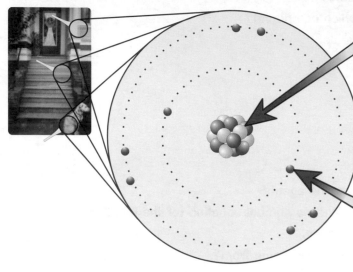

The Nucleus

1) The nucleus is in the <u>middle</u> of the atom.
2) It contains <u>protons</u> () and <u>neutrons</u> ( ).
3) Almost the <u>whole mass</u> of the atom is <u>concentrated</u> in the <u>nucleus</u>.
4) But size-wise it's <u>tiny</u> compared to the atom as a whole.

The Electrons

1) Electrons move <u>around</u> the nucleus in <u>shells</u>.
2) They're <u>tiny</u>, but their paths cover a <u>lot of space</u>.
3) Electron shells explain the <u>whole of Chemistry</u>.

You can't really look at an atom with a magnifying glass — they're way too small. You could fit about 10 million atoms across the full stop at the end of this sentence.

The Atomic Number is Just the Number of Protons

All atoms look pretty much like the picture above.
The only thing that changes is the <u>number</u> of protons, neutrons and electrons.

1) It's the number of <u>protons</u> in the nucleus that decides what <u>type</u> of atom it is.
2) For example, an atom with <u>1 proton</u> in its nucleus is <u>hydrogen</u>, an atom with <u>2 protons</u> is <u>helium</u>, an atom with <u>8 protons</u> is <u>oxygen</u>, an atom with <u>79 protons</u> is <u>gold</u>... and so on and so on.
3) The number of protons in an atom is called the <u>atomic number</u>.

The oxygen in the bottle and the gold bar are both elements — see page 52.

These bottles contain 21% <u>oxygen</u> atoms. <u>All</u> the oxygen atoms contain <u>8 protons</u>, so they all have an <u>atomic number</u> of <u>8</u>.

These are bars of <u>pure gold</u> — they contain only atoms of <u>gold</u>. <u>All</u> the atoms contain <u>79 protons</u>. And <u>all</u> the atoms have an <u>atomic number</u> of <u>79</u>.

Basic atom facts — they don't take up much space...

Two main points on this page...
1) If you look <u>really</u> closely, all atoms look roughly the same — with a cluster of protons and neutrons in the middle, and electrons whizzing round the outside.
2) Knowing an atom's <u>atomic number</u> is as good as knowing <u>what kind</u> of atom it is, and vice versa.

Chemical Symbols

Writing out the names of chemicals all the time can be a bit of a drag. That's why scientists invented <u>chemical symbols</u> — these let you write down <u>every</u> element using only <u>one or two letters</u>. Beautiful.

Atoms Can be Represented by Symbols

Atoms of each element can be represented by a <u>one or two letter symbol</u> — it's a type of <u>shorthand</u> that saves you the bother of having to write the full name of the element.

Some make <u>perfect sense</u>, e.g.

C = carbon   O = oxygen   Mg = magnesium

Others seem to make about as much sense as an apple with a handle, e.g.

Na = sodium   Fe = iron   Pb = lead

Most of these odd symbols actually come from the Latin names of the elements.

A Formula Shows What Atoms are in a Compound

<u>Compounds</u>, remember, are substances that contain <u>more than one kind</u> of atom.

1) <u>Carbon dioxide</u> is a <u>compound</u> — it contains carbon atoms and oxygen atoms.

2) In fact, every molecule of carbon dioxide contains <u>1 carbon atom</u> and <u>2 oxygen atoms</u>.

There you go... a molecule of carbon dioxide — one carbon atom and two oxygen atoms.

3) Using <u>chemical symbols</u>, this would be:

4) Easy.

CO_2

An atom of carbon...   *...and two atoms of oxygen.*

Chemical Equations Show the Atoms Involved in Reactions

1) A <u>chemical reaction</u> is when atoms are 'shuffled around'.

2) For example, when you <u>burn</u> a lump of <u>carbon</u>, a chemical reaction takes place. The carbon combines with <u>oxygen</u> from the air to form <u>carbon dioxide</u>:

An atom of carbon...   *...combines with two atoms of oxygen...*   *...and forms a molecule of carbon dioxide.*

3) A <u>chemical equation</u> just shows the same information about a reaction — '<u>what goes in</u>' and '<u>what comes out</u>'.

$$C + O_2 \rightarrow CO_2$$

Oxygen atoms always go round in pairs. So an oxygen molecule contains two oxygen atoms joined together — and you write O_2.

Don't panic — it's only a bunch of letters and numbers...

And another good thing is that these chemical formulas are international. So if you can get your head round what's going on here, then you'll be okay in your French exam if they suddenly start asking you about Chemistry questions. I know it's unlikely, but it's best to be prepared, I say.

Chemical Equations

During chemical reactions, things <u>don't</u> appear out of nowhere and things <u>don't</u> just disappear. You still have the <u>same atoms</u> at the <u>end</u> of a chemical reaction as you had at the <u>start</u> — they're just <u>arranged</u> in different ways.

Atoms Aren't Lost or Made in Chemical Reactions

1) The gas that a <u>kitchen cooker</u> burns is <u>methane</u>.

2) The <u>chemical formula</u> for methane is CH_4 — that's <u>one atom</u> of <u>carbon</u> (C) and <u>4 atoms</u> of <u>hydrogen</u> (H) in each molecule.

3) When methane burns, it reacts with <u>oxygen</u> from the <u>air</u> and you end up with <u>carbon dioxide</u> (CO_2) and <u>water</u> (H_2O).

$$CH_4 + O_2 \rightarrow CO_2 + H_2O$$

These equations aren't right however because they aren't balanced.

4) <u>BUT</u>... there must always be the <u>same</u> number of atoms of each element on <u>both sides</u> of the equation — they can't just <u>disappear</u>, and they can't just appear <u>out of nowhere</u>.

In this equation, the formulas are all okay but the <u>numbers</u> of the different kinds of atoms don't all <u>match up</u> on both sides. So that's not right.

5) You <u>can't</u> change formulas like H_2O to H_3O — all you can do is put numbers <u>in front</u> of the formulas where something doesn't balance.

Method: The Trick is to Balance Just ONE Type of Atom at a Time

The more you practise, the quicker you get, but all you do is this...

1) Find an element that <u>doesn't balance</u> and <u>pencil in a number</u> to try and sort it out.

2) <u>See where it gets you.</u> It may create <u>another imbalance</u> — if so, just pencil in <u>another number</u> and see where that gets you.

3) Carry on chasing <u>unbalanced</u> elements and it'll <u>sort itself out</u> pretty quickly.

<u>I'll show you.</u> In the equation above you're short of O atoms on the LHS (Left-Hand Side).

1) The only thing you can do about that is make it <u>$2O_2$</u> instead of just O_2:

$$CH_4 + 2O_2 \rightarrow CO_2 + H_2O$$

2) But that now causes <u>too many</u> O atoms on the <u>LHS</u>, so to balance that up you could try putting <u>$2H_2O$</u> on the RHS (Right-Hand Side):

$$CH_4 + 2O_2 \rightarrow CO_2 + 2H_2O$$

3) And suddenly there it is! <u>Everything balances.</u> There's 1 C, 4 H's and 4 O's on both sides.

$$CH_4 + 2O_2 \rightarrow CO_2 + 2H_2O$$

This is the <u>balanced symbol equation</u> that describes what goes on in a gas oven.

Balancing equations — weigh it up in your mind...

Remember what the different numbers mean... A number <u>in front of</u> a formula applies to the <u>entire</u> formula. So, <u>3</u>CH_4 means three lots of CH_4. The <u>little</u> numbers in the <u>middle</u> or at the <u>end</u> of a formula <u>only</u> apply to the atom <u>immediately before</u>. So the 4 in CH_4 just means 4 H's, not 4 C's.

Chemical Bonding and Properties

Atoms in an element or a compound aren't just held together by magic — they're held together by <u>chemical bonds</u>. The <u>properties</u> of the material are determined by the <u>type</u> of bonds in the compound.

Metal Properties are All Due to Metallic Bonds

1) Metals consist of a <u>giant structure</u> of atoms held together with <u>metallic bonds</u>.

2) These bonds are <u>really strong</u>.

3) Some of the <u>electrons</u> in each atom can <u>move about freely</u>. This creates a "<u>sea</u>" of <u>free electrons</u> throughout the metal.

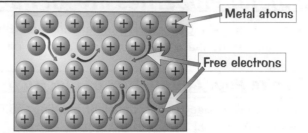

Metal atoms

Free electrons

1) They're Good Conductors of Electricity and Heat

The <u>free electrons</u> carry electrical current through the material, so metals are good conductors of <u>electricity</u>. They're also good at conducting heat. This makes metals great for electrical wiring and cooking pans.

2) Most Metals are Malleable

The layers of atoms in a metal can <u>slide</u> over each other, making metals <u>malleable</u> — they can be <u>hammered</u> or <u>rolled</u> into <u>flat sheets</u> or <u>pipes</u> (e.g. for plumbing).

There's more on p.75 about how metal properties determine what they're used for.

3) They Generally Have High Melting and Boiling Points

Metallic bonds are <u>very strong</u>, so it takes a lot of <u>energy</u> to break them — you have to get the metal <u>pretty hot</u> (except for mercury, which is a bit weird), e.g. copper melts at 1085 °C. This means your pan <u>won't melt</u> when you're cooking.

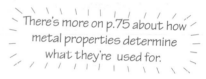

4) Some Metals are Hard

The <u>hardness</u> of a metal is a measure of how easy it is to <u>dent</u> it. Some metals (like iron) are hard, while some (like sodium) are <u>quite soft</u>. The more malleable a metal is, the less hard it is — makes sense really.

5) Some Metals are Dense

Density is all to do with how much <u>stuff</u> there is squeezed into a certain <u>space</u>. Metals feel <u>heavy</u> for their <u>size</u> (i.e. they're <u>very dense</u>) because they have a lot of <u>atoms</u> packed into a <u>small volume</u>.

6) Metals are Shiny when Polished or Freshly Cut

Metals reflect light from their smooth surfaces. This makes them look nice and <u>shiny</u>, which is great for things like posh kitchen appliances, e.g. kettles.

Free electrons — they don't cost you a penny...

It's all about the free electrons... because they can move the metal can conduct electricity (because electricity is just electrons flowing) and heat (because the electrons carry some of the heat along the metal). We'd be pretty stuffed without them... no more fry-ups, or electricity in the home for that matter.

Chemical Bonding and Properties

A non-metal is an <u>element</u> that's not a metal, e.g. oxygen and hydrogen.
Bonding <u>within</u> non-metals is different to bonding within metals...

Non-Metal Properties are All Due to Covalent Bonding

Non-metal atoms make <u>COVALENT</u> bonds with each other by <u>sharing electrons</u> with other atoms,
e.g. an oxygen atom shares electrons with another oxygen atom to make O_2 (oxygen gas).

1) They're Poor Conductors of Electricity and Heat

There are <u>no free electrons</u> in covalent bonding, so non-metals <u>don't usually conduct electricity</u> or <u>heat</u>.
This makes them rubbish for cooking pans but <u>great heat insulators</u>.

2) They Have Low Density and are Dull in Appearance

The non-metals that are <u>gases</u> have a <u>very low density</u> (obviously), and those that are <u>solids</u> or <u>liquid</u>
also have a <u>low density</u>. The solids are <u>dull</u> in appearance because they don't reflect light very well.

3) Those that Form Small Molecules Have Low Boiling and Melting Points

1) Some non-metals form <u>small molecules</u> by covalent bonding.

2) However, the <u>forces of attraction BETWEEN</u> these molecules
are <u>very weak</u>.

3) The <u>result</u> of these <u>feeble forces</u> is that the <u>melting</u> and
<u>boiling points</u> are <u>very low</u>, because the molecules are
<u>easily parted</u> from each other.

4) E.g. oxygen boils at −183 °C, which is why it's a <u>gas</u>
at room temperature.

Very weak forces
between the molecules

<u>Oxygen</u>

4) Those that Form Giant Structures Have High Boiling and Melting Points

1) Covalent bonding can also form <u>giant structures</u> — the atoms are <u>arranged</u> in a <u>regular lattice</u>.

2) The bonds are <u>very strong</u> and a lot of heat energy is needed to break the bonds,
giving them <u>high boiling</u> and <u>melting points</u>.

3) E.g. <u>diamond</u> is a giant <u>covalent</u> structure — it's just made of <u>carbon</u> and melts at 3550 °C.

Bonding Within Compounds can be Either Covalent or Ionic

A compound contains <u>different atoms</u> bonded together, e.g. H_2O or NaCl.
The <u>same bonding rules</u> apply to compounds as to non-metals:

1) A <u>small covalently</u> bonded compound (e.g. <u>water</u>, <u>methane</u> or
<u>carbon dioxide</u>) will have a <u>low boiling</u> and <u>melting point</u>.

2) <u>Giant</u> compounds have <u>high boiling</u> and <u>melting points</u>,
Example 1: <u>silica</u> is a giant compound formed by <u>covalent</u> bonding.
It's made of silicon and oxygen atoms and melts at 1710 °C.

Example 2: <u>salt</u> is a giant compound formed by <u>ionic</u> bonding.
It's made of sodium and chlorine ions and melts at 801 °C.

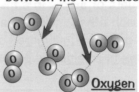

In <u>ionic bonding</u>, atoms <u>lose</u>
<u>or gain electrons</u> to form
<u>charged particles</u> (ions),
which are then <u>strongly</u>
<u>attracted</u> to one another.

Giant covalent structures — a girl's best friend...

OK, so bonding <u>within</u> a <u>metal</u> is always <u>metallic bonding</u>. Bonding <u>within</u> a <u>non-metal element</u> is
<u>covalent</u> and bonding within any type of <u>compound</u> can be <u>either</u> covalent or ionic. Got that? Good.

Checklist: Symbols and Formulas

OK, I'm warning you now that this page might be a bit dull. Unfortunately, you still need to learn it. For your exam you need to know the <u>chemical symbols</u> for some <u>molecules</u> and some <u>elements</u>. For the elements you also need to know whether they are <u>metals</u> or <u>non-metals</u>... oh the excitement.

Some Elements are Metals and Some are Non-metals

Metals

There are some that you'll <u>know</u> are metals...

Aluminium **Al**   Gold **Au**   Iron **Fe**

Silver **Ag**   Lead **Pb**   Zinc **Zn**

...and some less obvious ones.

Calcium **Ca**   Magnesium **Mg**

Potassium **K**   Sodium **Na**

Non-metals

There are some that you'll know are definitely <u>not metals</u> because they're <u>gases</u>...

Oxygen **O**   Nitrogen **N**   Hydrogen **H**

...and some you'll just have to learn.

Argon **Ar**   Bromine **Br**   Sulfur **S**   Carbon **C**

Chlorine **Cl**   Phosphorus **P**   Silicon **Si**

You Need to Know the Formulas of These Molecules...

Remember that molecules contain <u>more than one</u> atom. Here are the ones you need to learn:

These ones are gases at room temperature...

Oxygen O_2 Methane CH_4 Hydrogen H_2 Carbon dioxide CO_2

Nitrogen N_2 Chlorine Cl_2 Hydrogen chloride HCl Ammonia NH_3

...and these ones are liquid:

Water H_2O Ethanol C_2H_5OH

Dissolved in water this is hydrochloric acid.

...and How to Work Out the Formula for Ionic Compounds

You learnt about <u>compounds</u> and <u>ionic bonding</u> on p. 70 — it can be difficult to remember these formulas, but it's not too hard to work them out. Atoms can form <u>ions</u> when they have <u>lost</u> or <u>gained</u> an <u>electron</u> — so they have either a <u>positive</u> or a <u>negative charge</u>. The main thing to remember is that in compounds where these atoms are bonded the total charge must always add up to <u>zero</u>, so you just need to balance out the charges. For example:

EXAMPLE: Find the formula for <u>zinc carbonate</u>.

Find the charges on a zinc ion and a carbonate ion. A zinc ion is Zn^{2+} and a carbonate ion is CO_3^{2-}. To balance the total charge you need one zinc ion to every one carbonate ion. So the formula of zinc carbonate must be:

$$ZnCO_3$$

EXAMPLE: Find the formula for <u>aluminium oxide</u>.

Find the charges on an aluminium ion and an oxide ion. An aluminium ion is Al^{3+} and an oxide ion is O^{2-}. To balance the total charge you need two aluminium ions to every three oxide ions. So the formula of aluminium oxide must be:

$$Al_2O_3$$

H_2O, CO_2, Au, Ar, DVD, FBI, GSOH...

Unfortunately there's no way round <u>just learning</u> this stuff — close the book and <u>scribble out</u> the 10 metals, the 10 non-metals and the 10 compounds on this page. If you only get nine for each category then have another read of the page and <u>try again</u> till you get all 30.

Revision Summary for Section 2.6

That was a pretty theoretical section, but all that theory helps you understand things like why metals conduct electricity, and why different molecules have different properties. Make sure you learn it — it'll make the rest of the chemistry a bit easier. You know the drill by now — do the questions and if there are any you can't do go back and read the page again till that big sponge in between your ears is full with chemistry basics...

1) Name the three particles that make up an atom.

2) Which particles make up the nucleus?

3) Which particles move around the nucleus in shells?

4) The number of which particle decides what type of atom it is?

5)* Copper has 29 protons. What is its atomic number?

6)* Carbon has an atomic number of 6. How many protons does it have?

7) What are chemical symbols?

8) What is a compound?

9)* How many hydrogen atoms does methane have?

10)* How many carbon atoms does ethanol have?

11) What does a chemical equation show?

12)* Balance the following symbol equations:

a) $Mg + O_2 \rightarrow MgO$

b) $Cl_2 + KBr \rightarrow Br_2 + KCl$

c) $C_6H_{12}O_6 \rightarrow C_2H_5OH + CO_2$

d) $Na + H_2O \rightarrow NaOH + H_2$

13) What type of bonding occurs in metals?

14) Describe how the type of bonding makes metals good conductors of electricity.

15) Why are some metals malleable?

16) Why do most metals have high melting and boiling points?

17) Give two other properties of metals.

18) Describe what happens to the electrons in an atom when covalent bonds are formed.

19) Describe what happens to the electrons in an atom when ionic bonds are formed.

20) Explain why non-metals are poor conductors of electricity.

21) Why do small molecules with covalent bonds have low boiling and melting points?

22) Why do giant structures have high boiling and melting points?

23) Give an example of one giant ionic structure and one giant covalent structure.

24) Is potassium a metal or a non-metal? What is its chemical symbol?

25) Is silicon a metal or a non-metal? What is its chemical symbol?

26) Is calcium a metal or a non-metal? What is its chemical symbol?

27) What is the formula for ammonia?

28) What is the formula for ethanol?

29) What is the formula for nitrogen gas?

30) What is the formula for water?

31) What does the charge of a compound always add up to?

* Answers on page 148.

Limestone

Limestone is a rock that's used for building stuff like houses and churches — see page 53 for a bit more. You need limestone to make mortar, cement, concrete and glass too. In fact, it's blooming marvellous.

Limestone is Used as a Building Material

1) Limestone is a bit of a boring grey or white colour. It's formed over thousands of years, often from sea shells.

2) It's quarried out of the ground. This causes some environmental problems though — see next page.

3) It's great for making into blocks for building with. Fine old buildings like cathedrals are often made purely from limestone blocks. It's also used for statues and fancy carved bits on nice buildings too.

4) Limestone can also be crushed up into chippings and used in road surfacing.

The Parthenon in Greece is made from limestone.

Limestone is Mainly Calcium Carbonate

1) Limestone is mainly calcium carbonate — $CaCO_3$.

2) When it's heated it breaks down to make calcium oxide (quicklime) and carbon dioxide.

> calcium carbonate → calcium oxide + carbon dioxide
> (limestone)   (quicklime)
> $$CaCO_3 \rightarrow CaO + CO_2$$

Quicklime is mainly produced to make slaked lime.

This type of reaction is known as an endothermic reaction because it needs heat to happen:

> An **ENDOTHERMIC REACTION** is one which **TAKES IN ENERGY** from the surroundings, usually in the form of **HEAT**, which is shown by a **FALL IN TEMPERATURE**.

Quicklime Reacts with Water to Produce Slaked Lime

1) When you add water to quicklime you get slaked lime. Slaked lime is actually calcium hydroxide.

> calcium oxide + water → calcium hydroxide
> (quicklime)   (slaked lime)
> $$CaO + H_2O \rightarrow Ca(OH)_2$$

2) Slaked lime is an alkali which can be used to neutralise acid soils in fields. Powdered limestone can be used for this too, but the advantage of slaked lime is that it works much faster.

3) This type of reaction is known as an exothermic reaction because it produces heat.

> An **EXOTHERMIC REACTION** is one which **GIVES OUT ENERGY** to the surroundings, usually in the form of **HEAT**, which is shown by a **RISE IN TEMPERATURE**.

Limestone — a sea creature's cemetery...

Building with limestone sounds like the best thing since sliced bread — but it's not so great when faced with acid rain. The acid reacts with the limestone and dissolves it away, leaving you with an unsightly building.

Limestone

Limestone can be <u>combined</u> with <u>other materials</u> to make different <u>building materials</u>. But using limestone ain't all hunky-dory — tearing it out of the ground and making stuff from it causes quite a <u>few problems</u>.

Limestone is Used to Make Other Useful Building Materials

1) Powdered limestone is <u>heated</u> in a kiln with <u>powdered clay</u> to make <u>cement</u>.

2) Cement can be mixed with <u>sand</u> and <u>water</u> to make <u>mortar</u>. <u>Mortar</u> is the stuff you stick <u>bricks</u> together with.

3) Or you can mix cement with <u>sand</u>, <u>water</u> and <u>gravel</u> to make <u>concrete</u>.

4) And believe it or not limestone is also used to make <u>glass</u>. You just heat it with <u>sand</u> and <u>sodium carbonate</u> until it melts.

5) Cement, mortar, concrete and glass are all <u>building materials</u>.

Producing Limestone Products Damages the Environment

<u>Digging up</u> and <u>processing</u> limestone (to make cement and glass etc.) can cause <u>environmental problems</u>...

1) For a start, it makes <u>huge ugly holes</u> that permanently change the landscape.

2) <u>Quarrying</u> processes, like blasting rocks apart with explosives, make lots of <u>noise</u> and <u>dust</u> in quiet, scenic areas.

3) Quarrying <u>destroys the habitats</u> of animals and plants.

4) The limestone needs to be <u>transported away</u> from the quarry — usually in lorries. This causes more noise and pollution.

5) Waste materials produce <u>unsightly tips</u>.

6) Cement factories make a lot of <u>dust</u>, which can cause <u>breathing problems</u> for some people.

7) <u>Energy</u> is needed to produce cement and quicklime. The energy is likely to come from burning <u>fossil fuels</u>, which causes pollution (see p.63).

But on the Plus Side...

Even though producing limestone materials has some damaging effects, the industry also has its <u>advantages</u>:

1) Limestone provides things that people want — like <u>houses</u> and <u>roads</u>. Chemicals used in making <u>dyes</u>, <u>paints</u> and <u>medicines</u> also come from limestone.

2) Limestone products are used to <u>neutralise acidic soil</u> and <u>acidic water</u>.

3) Quicklime (from limestone) is used in power station chimneys to <u>neutralise sulfur dioxide</u>, which is a cause of acid rain (see p.63).

4) The quarry and associated businesses provide <u>jobs</u> for people and bring more money into the <u>local economy</u>.

5) Once quarrying is complete, <u>landscaping</u> and <u>restoration</u> of the area is normally required.

Tough revision here — this stuff's rock hard...

There's a <u>downside</u> to everything, including using limestone — ripping open huge quarries can <u>spoil the countryside</u>. In the exam you might have to <u>evaluate</u> the effects of producing limestone materials. Make sure you don't <u>just</u> go on about how bad mining limestone is for the environment — there are <u>economic effects</u> (things to do with money) and <u>social effects</u> (the effect on people in the area) too.

Metals and Alloys

We can get a lot of <u>useful materials</u> from the <u>surface</u> of our <u>little planet</u>. Not only fancy rocks but also <u>metals</u> (see p.55). Different metals are used for different things in construction, depending on their <u>properties</u>. Pure metals often aren't quite right for certain jobs. Instead of just making do, metals can be <u>mixed</u> with other metals (or non-metals) to make <u>more suitable materials</u> — these mixtures are called <u>alloys</u>.

Different Metals and Alloys have Different Properties

Lead

1) <u>Soft</u> and <u>malleable</u> — it can be beaten flat and moulded into whatever shape you need.
2) <u>Unreactive</u> — it doesn't corrode.

Lead is moulded around the <u>joins</u> in a <u>roof</u> (e.g. where the chimney joins on) to form a seal that stops <u>water</u> getting in — this is called <u>flashing</u>.

Iron / Steel

1) <u>High tensile strength</u> — it can take a lot of weight before it bends or breaks.
2) <u>Cheap</u> to produce.
3) <u>Rusts</u> very easily.

Steel is used for its strength as a <u>support material</u>.

Aluminium

1) <u>Lightweight</u>.
2) <u>Resistant</u> to <u>corrosion</u> — it doesn't wear away.

Aluminium is used for <u>window frames</u>.

LEAD "flashing" directs <u>water</u> away from <u>joins</u> in the roof.

<u>ALUMINIUM</u> is used in <u>window frames</u>.

Brass

Brass is an <u>alloy</u> containing <u>copper</u> and <u>zinc</u>.

1) <u>Malleable</u> — can easily be <u>formed</u> into any shape you want.
2) Quite <u>resistant</u> to <u>tarnishing</u> — not easy to mark.

Brass is great for <u>fancy things</u> like <u>handles</u>, <u>hinges</u> and <u>taps</u>.

In modern buildings, <u>IRON</u> or <u>STEEL</u> girders are used to <u>support</u> roofs and upper floors.

<u>COPPER</u> wires carry <u>electricity</u> around the house for <u>lighting</u> and electrical <u>appliances</u>.

<u>SOLDER</u> is an alloy that is melted and used to <u>join</u> wires or pipes.

<u>COPPER</u> is used to make <u>hot water tanks</u> and <u>water pipes</u>.

<u>BRASS</u> is used to make <u>fixtures</u> and <u>fittings</u>.

Solder

Solder is an <u>alloy</u>, usually containing <u>tin</u> and <u>lead</u>.

1) Relatively <u>low melting point</u> — so it can be melted to <u>join wires</u> or <u>pipes</u>.
2) <u>Conducts electricity</u>.

Solder is great for <u>connecting electric wiring</u> and <u>plumbing pipes</u>.

Copper

1) <u>Malleable and ductile</u> — it can be beaten into sheets, rolled up to make <u>pipes</u> and can be drawn out into <u>long wires</u>.
2) <u>Excellent conductor of electricity</u>.
3) <u>High melting point</u> — so it can cope well with high temperatures.
4) <u>Doesn't corrode</u> with water.

Copper is ideal for <u>water pipes</u> and <u>hot water tanks</u>. It doesn't <u>corrode</u> like iron, or <u>melt</u> like some plastics. Very pure copper is used to make the wires in <u>electrical circuits</u>.

Our house — in the middle of our street...

You wouldn't want to use steel to seal gaps in your roof. A couple of decent rain showers and you'd just have a pretty little pile of rust and a wet carpet. You need to <u>match uses</u> to <u>properties</u>.

Polymers

The plastics we use in day-to-day life are all polymers. Plastics are amazingly useful things — you can keep things in them, wear them, carry things in them or hide under them. They come in lots of different forms...

Polymers' Properties Decide What They're Used For

Different polymers have different physical properties — some are stronger, some are stretchier, some are more easily moulded, and so on. These different physical properties make them suited for different uses.

1) __FLEXIBILITY__ — some polymers are pretty bendy and stretchy. Polyethene is a flexible polymer which is used to make plastic bags and squeezy bottles. PVC can be used to make synthetic leather, which is pretty stretchy.

2) __BEHAVIOUR ON HEATING__ — some polymers have a low melting point so they're no good for anything that'll get very hot, e.g. a kettle. However some polymers are heat-resistant, e.g. polypropene can be used to make plastic kettles. When melted, polymers can easily be moulded into different shapes.

3) __POOR CONDUCTORS OF HEAT__ — polymers generally don't transfer heat well — they're good insulators, e.g. they're great for things like coffee cups and pan handles. (Polymers are great for pan handles because they don't allow the heat from the pan to get to your hands.)

4) __POOR CONDUCTORS OF ELECTRICITY__ — polymers don't allow an electric current to flow through them.
So, handily, they can be used as electric wire insulation — the electricity is trapped in and can only flow down the wire, which makes them safe for you to handle. The casing of many appliances (e.g. hairdriers) are plastic for the same reason.

If you're making a product, you need to pick your polymer carefully. It's no good trying to make a kettle out of a plastic that melts at 50 °C — you'll end up with a messy kitchen, a burnt hand and no cuppa. You'd also have a bit of difficulty trying to wear clothes made of brittle, un-bendy plastic. So, think about what job you need a plastic to do and what properties it needs to do it.

Plastic pans — not such a great idea...

Polymers sound great but they're pretty hard things to get rid of — they aren't biodegradable, so they don't rot. This can be useful until you need to get rid of your plastic. Burning them is damaging to the environment and if you throw them away they end up in landfill sites — and could be there for thousands of years. So, the best thing to do is reuse and recycle. Some clever scientists have invented a biodegradable plastic bag. But this doesn't solve the problem of all the millions already hanging around on our planet.

Ceramics

People have been making and using ceramics for <u>thousands</u> of years, e.g. <u>pottery</u> and <u>glass</u> are ceramics. Today ceramics are used loads in <u>construction</u>.

Ceramics are Materials Made by Heating

1) <u>POTTERY</u> and <u>PORCELAIN</u> — made by <u>heating clay</u>. When it's <u>wet</u> clay can be <u>moulded</u> into any shape you like. 'Firing' it in a hot oven (a kiln) makes it turn <u>hard</u> and keep its <u>shape</u>. They can be <u>glazed</u> to add colours or patterns.

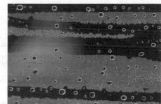

2) <u>GLASS</u> — made by <u>heating sand</u> with other chemicals. It's <u>transparent</u>, and can be made <u>different colours</u>. Glass can be made into <u>any shape</u>, including flat sheets.

3) <u>CEMENT</u> — made by <u>heating limestone</u> with <u>clay</u>. When you mix cement with water it becomes runny and then sets hard. Cement is used to <u>stick bricks</u> and <u>stone</u> together, and to make <u>concrete</u>.

Ceramics are Hard-Wearing, but Brittle

<u>USEFUL PROPERTIES OF CERAMICS:</u>
1) <u>Hard-wearing</u> — they don't scratch easily.
2) <u>High melting point</u> — they are <u>heat-</u> and <u>fire-resistant</u>.
3) <u>Waterproof</u> and <u>smooth</u> — easily <u>cleaned</u> and <u>hygienic</u>.
4) <u>Inert</u> — they don't <u>corrode</u> or <u>react</u> with chemicals.
5) <u>Electrical insulators</u> — ceramics <u>don't conduct electricity</u>.
6) <u>Attractive appearance</u> — available in <u>different colours</u> and <u>patterns</u>.

<u>A NOT-SO-USEFUL PROPERTY:</u>
They're <u>brittle</u> — they aren't very <u>flexible</u> and <u>break easily</u>.

They're Useful for Building Houses and Serving Food

These properties make ceramics useful for loads of building materials and household things...

<u>WINDOWS</u> — transparent and hard-wearing.

<u>BRICKS</u> — cement used to stick bricks together.

<u>FLOOR / WALL TILES</u> — easy to clean and attractive.

<u>TOILETS AND SINKS</u> — waterproof and don't react with chemicals.

<u>CUPS AND PLATES</u> — insulate heat.

<u>SPARK PLUGS</u> in engines — heat-resistant electrical insulators.

So be careful with your mum's best china...

...it smashes easily, but more importantly she'll probably kill you.

Ceramics don't burn, melt, rot, rust or get eaten by termites, which is handy really. The only downside is they're <u>easily broken</u> or smashed — so keep your cricket balls and clumsy hands away from them.

Composites

Manufacturers and designers can't always find one material that has all the properties they want. A mixture of <u>two materials</u> often provides the solution — these are called <u>composites</u>.

Composites Combine the Properties of Two or More Materials

There are two main types of composites:

1) FIBRE-REINFORCED COMPOSITES

These are the <u>most commonly</u> used composites. One material provides <u>flexibility</u> and the other provides the <u>strength</u>. Useful properties include:

1) Good <u>strength/weight</u> ratio — strong compared to their weight.
2) <u>Resistant to corrosion</u> — won't wear away easily.

Here are some examples of fibre-reinforced composites:

GRP — Glass-reinforced plastic
GRP is <u>plastic reinforced</u> by <u>small fibres</u> of <u>glass</u>.
This makes it <u>stronger</u> than <u>plastic</u> but not as <u>brittle</u> as <u>glass</u>. It can also be <u>moulded</u> into different shapes. It's used to make things like <u>baths</u>.

Reinforced concrete
This is <u>concrete</u> made <u>stronger</u> by adding <u>fibres</u> or <u>rods</u> of <u>metal</u> — usually <u>steel</u>. This is used in the construction of <u>buildings</u>.

Laminated and reinforced glass
Glass can be made <u>stronger</u> in two ways:
1) by adding a <u>metal grate</u> to it — this is called <u>reinforced glass</u>.
2) by <u>sticking two sheets</u> of <u>glass</u> <u>together</u> — this is called <u>laminated glass</u>.
Both are <u>stronger</u> than normal glass and are used in <u>security doors</u> and <u>windows</u>.

2) PARTICLE-BASED COMPOSITES

These are <u>mixtures</u> of '<u>bits</u>' of one substance in a <u>paste</u> or resin of another.

Useful properties include:
1) Very <u>strong</u> when <u>compressed</u> — don't squash easily under heavy weights.
2) <u>Strong throughout</u> — due to its <u>regular</u> <u>structure</u> there are no weaker bits.

Cermet
<u>Ceramic</u> (cer) and <u>metal</u> (met) can be mixed together to make <u>cermet</u>. This is a <u>heat-resistant</u> but <u>mouldable material</u>.
It's used in <u>electric</u> <u>components</u> and <u>solar</u> <u>water heaters</u>.

Concrete
This is a mixture of <u>cement</u>, <u>water</u>, <u>sand</u> and <u>stone</u>.
It's <u>strong</u> and can be <u>moulded</u> into complex shapes. It's used for <u>roads</u> and <u>buildings</u>.

Plywood / MDF
These are materials made out of bits of <u>wood</u> <u>stuck together</u> with <u>glue</u>. Plywood is <u>thin sheets</u> of wood <u>stacked</u> together. MDF (medium-density fibreboard) is <u>sawdust</u> stuck together.
They're used loads in <u>furniture</u> and <u>flooring</u>.

Concrete is a particle-based composite but when it's reinforced with another material it can be classed as a fibre-reinforced composite.

In some cases two things are better than one...

Don't forget that the <u>properties</u> of <u>composites</u> come from the properties of their <u>components</u>, e.g. mint choc chip ice cream is so tasty because it combines crunchy chocolate bits with minty fresh ice cream. Mmmm.

Choosing Materials for a Product

With so many different materials to choose from it's hard to decide which one is best for a job.

Knowing the Properties of Materials Will Help You Decide

Here are some of the good (and not so good) properties of manufactured materials:

Material	Advantages	Disadvantages
Metals	Strong and hard Conduct heat and electricity	Some are heavy (e.g. steel) Some are expensive (e.g. gold) Some corrode (e.g. iron)
Polymers	Cheap and light	Non-renewable Non-biodegradable
Ceramics	Hard-wearing Insulate heat and electricity	Brittle
Composites	Strong Flexible	Some are non-renewable Some are non-biodegradable

Think about these properties when choosing a material for a product.

You can Choose from Modern or Traditional Building Materials

1) When constructing a building you can choose from traditional or more modern building materials. Traditional building materials tend to be naturally occurring, e.g. wood and stone, and modern ones are usually man-made, e.g. polymers, ceramics, composites and some metal alloys.

2) If you're asked in the exam to pick between them don't just say, 'Modern materials are always way better than traditional ones'. Modern materials have advantages over traditional materials, but they also have disadvantages. Here are a couple of examples:

Plastic or wood window frames?
Plastic is more durable than wood and it's resistant to corrosion. But it's also non-biodegradable and often doesn't look as nice as wooden frames.

Steel beams or wood beams?
Steel beams are stronger than wooden ones but they're non-biodegradable and they don't look as nice (if they're exposed that is).

Plywood or wood panelling?
Plywood is cheaper than wood but it's easily warped (changes shape) and doesn't look as nice.

Reinforced glass or normal glass windows?
Reinforced glass is stronger so more resistant to breakage, but it's also more expensive than normal glass.

3) So, whether you use traditional, natural materials or modern, man-made materials depends on what properties you want. E.g. if you're rebuilding an old farmhouse you'll probably use wood beams to match the exposed beams already there. But if you're building a block of high-rise flats you'll use steel beams because they're stronger and no one will see them (so they don't have to look nice).

More than one type of material may be suitable for a product...

What's important is that you can give reasons to explain your choice. Use information you remember about the properties of materials and any information given to you in the exam to help you.

Choosing Materials for a Product

The materials you choose to use when making a product are very important — think carefully about which material is suited for the selected job, because if you don't your product might not work.

Consider the Product Before Choosing Your Materials

When you make a product (e.g. a shed or a block of flats) you need to make sure you know what properties are important. This information might be given to you in the form of a design specification — a list of conditions you need to meet in order for the product to work well. From this you can choose the materials that best fit your needs. Here's an example:

1) You need to design a garden shed to house a collection of gnomes during the winter. This is the design specification you're given.

> **DESIGN SPECIFICATION: GARDEN SHED FOR GNOMES**
>
> 1) Strong and sturdy (may need extra reinforcements)
> 2) Solid base / foundation
> 3) Waterproof (you don't want your precious gnomes getting wet)
> 4) Cheap
> 5) Transparent windows (your gnomes need a good view to stop them getting bored)
> 6) Secure roof
> 7) Entrance

2) There are loads of materials to choose from — metal, polymers, ceramics and composites. Think about what materials have the right properties.

3) Here are a few examples of the materials you might pick for the shed, along with the reasons for choosing them:

The roof needs to be waterproof — ceramics, metals and polymers are all waterproof but ceramics would be too heavy so a light metal or plastic roof would be perfect.

To help strengthen the structure you could use metal supports because metal is strong.

The windows need to be transparent and durable (to withstand the weather). Glass or clear plastic both fit the bill as they're both transparent and hard-wearing.

Fixtures and fittings (like hinges and handles) need to be durable and fairly resistant to corrosion. Plastic and brass have these properties so would be ideal.

The walls need to be strong, sturdy, cheap and durable — plywood would be a good choice because it's all of these things.

You need a strong base — concrete would be a good choice as it's strong and can be poured into any shape.

A paper shed — you should've gnome better...

So, find the properties you need to make your product work, then pick materials with the exact same properties — not too difficult really, if you use your common sense.

Revision Summary for Section 2.7

Just think, you only have to know this stuff for your exam but construction engineers have to know this stuff for their jobs. They're the people who figure out what's the best thing to build things out of — everything from garages to enormous hotels and bridges. Anyway, like I said, you need to know this stuff for your exams so quit dreaming of your gnome collection and try these questions:

1) What is limestone used for?
2) How is quicklime produced?
3) How is slaked lime produced?
4) What is slaked lime used for?
5) What type of reaction produces heat?
6) What type of reaction takes in heat?
7) Give four ways that digging or processing limestone can damage the environment.
8) Give three advantages of the limestone industry.
9) Name three alloys.
10) Give four properties of copper.
11) What is lead used for in construction?
12) What is steel used for in modern buildings?
13) Why is solder suitable for connecting electric wiring?
14) Why is polyethene good for making squeezy bottles?
15) Why are polymers suitable for pan handles?
16) Why are polymers used as electric wire insulators?
17) Name three types of ceramics.
18) What is glass made from?
19) Give four useful properties of ceramics.
20) Give an unwanted property of ceramics.
21) What are composites?
22) Name the two main types of composites.
23) Give two useful properties for each type of composite.
24) What is reinforced concrete usually used for?
25) What two materials are used to make cermet?
26) What material is added to glass to make reinforced glass?
27) Give an advantage and disadvantage of each type of material:
    a) metals
    b) polymers
    c) ceramics
    d) composites
28) Give one reason why steel beams are sometimes used instead of wood beams in construction.
29) Give one reason why you might pick a traditional building material rather than a modern building material.
30) What is a design specification?
31)* What material might you use for: a) a frying pan base, b) glass in a security door, c) a toilet, d) a coffee pot?

* Answers on page 148.

Energy in the Home

There are Lots of Different Forms of Energy:

...but energy is generally <u>only useful</u> when you can <u>convert</u> it from one form to another.
For example, electrical energy is very useful because we can use devices to convert it into
many useful forms of energy at the touch of a button:

- <u>LIGHT ENERGY</u>, e.g. light bulbs, **TVs**
- <u>HEAT ENERGY</u>, e.g. kettles, toasters, fan heaters
- <u>KINETIC (MOVING) ENERGY</u>, e.g. anything with an electric motor in it
- <u>GRAVITATIONAL POTENTIAL ENERGY</u>, e.g. anything that lifts, like a stairlift

Different Energy Sources are Useful for Different Things

Energy doesn't just come in the form of <u>mains electricity</u> — some homes use <u>different sources</u> of
energy, such as <u>natural gas</u>, <u>oil</u>, <u>wood</u> or <u>coal</u>. And then there's <u>battery power</u>, of course.
Each source contains <u>energy</u> that can be changed into a form of energy that we need.
You need to be able to <u>explain why</u> a particular energy source is chosen for a task:

1) **MAINS ELECTRICITY — VERY CONVENIENT:**
Mains electricity can provide enough power for <u>all your domestic appliances</u>,
even those that need <u>loads of power</u>, like dishwashers. The main drawback is
safety — the mains supply is at <u>high voltage</u> (230 V) and an electric shock
from the mains could kill you. It's also pretty <u>expensive</u>.

2) <u>NATURAL GAS — CHEAPER for HEATING</u>:
Mains gas is used in some homes for <u>heating</u> and
<u>cooking</u>. It's often <u>cheaper than electricity</u>. It burns
efficiently and has few impurities — it's a <u>clean fuel</u>.
But you can <u>only</u> use it to provide <u>heat</u> — it won't
power the TV, for example. You can also get natural
gas in <u>bottles</u> if there's no access to mains gas, but
it's a bit more expensive.

mains gas   bottled gas

3) <u>OIL — CONVENIENT for HEATING in REMOTE AREAS</u>:
This is used in some homes instead of gas for <u>heating</u>
and <u>cooking</u>. As it is a <u>liquid</u>, it's easier to store than
natural gas and is often used in <u>rural areas</u> where it's
<u>not possible</u> to lay <u>gas pipelines</u> to homes.

oil tank

4) <u>BATTERIES — LOW POWER but PORTABLE</u>:
These are a <u>portable alternative</u> to mains electricity for powering <u>electrical devices</u>. They're great
for things like <u>mobile phones</u> and <u>personal stereos</u>, where you don't
want to be plugged into the wall. They're also useful for things like
<u>shower radios</u>, where it might not be <u>safe</u> to have a mains socket.
But batteries are <u>more expensive</u> than mains, and they generally
produce <u>less power</u> (see next page), so, for example, you couldn't
realistically use them to run your central heating system.

There are lots of forms of energy — but cake's my favourite...

Remember — electricity is so useful around the home because it's so easy to transform it into other
forms of energy — if you have all the necessary gadgets, that is. Learn and enjoy.

Electrical Appliances

Most people in the UK use electricity every day — even if it's just for boiling the kettle and putting the lights on. Then there's washing machines, computers, lava lamps... In all those appliances, <u>electrical energy</u> is converted into other, <u>useful</u> forms of energy (OK, the usefulness of a lava lamp is debatable).

An Appliance Forms Part of a Circuit

When you plug an appliance into a wall socket, you're connecting it to the mains electricity supply:

1) Two of the 'pins' on the plug are now connected to two copper wires which run through the walls to the point where the electricity supply comes into your house (also along copper wires).

2) This makes an electrical <u>circuit</u> a bit like the one shown below. (It's not quite like this but you don't have to worry about the details.)

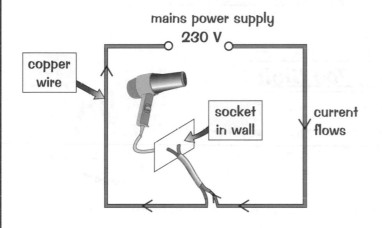

mains power supply
230 V
copper wire
socket in wall
current flows

3) When you switch the appliance on, all the bits of wire are connected up and an electric <u>current</u> flows round the circuit — and through your appliance. <u>Current</u> is measured in units called <u>amps</u> (or amperes), <u>A</u>.

4) Current is 'pushed' around the circuit by the power supply. This 'push' is called the <u>voltage</u> or the <u>potential difference</u> of the supply. <u>Voltage</u> is measured in <u>volts</u>, <u>V</u>.

The Power of a Circuit Depends on Voltage and Current

A circuit's <u>power</u> is how quickly it converts electrical energy into other forms (see p.86 for more on this). It's fairly easy to work out if you've got the right information about the circuit.

1) <u>Power</u> depends on the <u>voltage</u> of the supply and the size of the <u>current</u> flowing.

2) Power has units of <u>watts</u>, <u>W</u>.

3) You'll need to learn this formula and practise using it:

POWER = VOLTAGE × CURRENT
(watts) (volts) (amps)

There's a symbol version too — power is P, voltage is V and current is I (oddly). So:

$$P = V \times I$$

<u>EXAMPLE:</u>
James puts the kettle on.
The voltage of the mains supply is 230 V. The current flowing in the circuit is 9 A.
Calculate the power of this circuit.
<u>ANSWER:</u> P = V × I = 230 × 9 = 2070 W.

Circuit training — keep fit while you learn...

Mains electricity is great — you just plug in, switch on and there it is... it's almost magical. Never stick your magic wand in a socket, though. Instead, learn about power, voltage and current — much safer.

Electrical Safety

If an appliance connected to the mains supply goes wrong, you could find yourself as the 'appliance' in an electrical circuit. Very unpleasant. That's where fuses, earth wires and circuit breakers come in.

Most Electrical Cables Have Three Wires

1) Most cables have <u>three</u> wires — which connect to the <u>three pins</u> of a plug. Each wire is coated in plastic insulation.

2) If everything's working properly, current flows in and out of the <u>live</u> and <u>neutral</u> wires.

3) Normally, <u>no current</u> flows in the <u>earth</u> wire. It's just there for safety (see below).

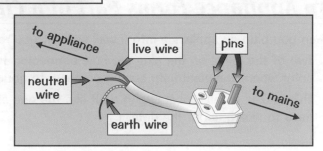

Fuses Melt When the Current is Too High

Most electrical appliances have a <u>fuse</u> in their circuits.

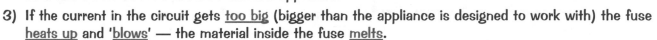

1) The fuse is a small component inside the plug. It's connected at either end to the <u>live</u> wire (which is connected to the incoming electricity supply).

2) Normally, the fuse is just part of the circuit — current passes through as it flows between the live wire and the appliance.

3) If the current in the circuit gets <u>too big</u> (bigger than the appliance is designed to work with) the fuse <u>heats up</u> and '<u>blows</u>' — the material inside the fuse <u>melts</u>.

4) This means that the circuit is now <u>broken</u> — the 'live' wire going to the appliance isn't connected to the mains supply any more. So the appliance is now completely cut off from the mains supply.

Earthing Prevents Electric Shocks

If an appliance develops a <u>fault</u> and the <u>live wire</u> somehow touches exposed metal on the appliance, someone who touched that metal part could get an <u>electric shock</u> — current could flow through their body, out of their feet and into the ground. That would be bad for them. <u>Earth wires</u> help prevent this:

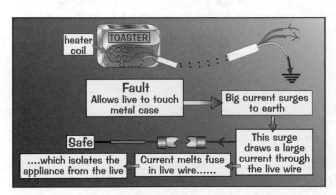

1) One end of the earth wire is connected to the exposed metal parts of the appliance.

2) The other end is connected to <u>the Earth</u> (often via the building's metal cold water pipes).

3) If exposed metal on the appliance becomes <u>live</u>, a big current flows in through the live wire and out via the <u>earth wire</u> to the ground.

4) This large current <u>blows the fuse</u> in the plug and <u>cuts off</u> the <u>live supply</u>. Someone touching the appliance would now be safe.

Toast the bread — not yourself...

All appliances with <u>exposed metal</u> (like toasters) must be "<u>earthed</u>" to reduce the danger of <u>electric shock</u>. "Earthing" just means attaching the metal parts to the earth wire in the cable. It's well worth doing.

Electrical Safety

With fuses, size matters. You need to know how to choose the <u>correct fuse</u> for a particular appliance.

Different Appliances Have Different Fuse Ratings

1) <u>Fuses</u> are designed to stop the current flowing through an appliance if it gets too high (see p.84). This <u>protects</u> the user from <u>electric shocks</u>. But it also prevents the appliance being damaged.

2) Fuses come in different <u>current ratings</u>: 3 amps (A), 5 A, 13 A, etc.

3) You should always use a fuse that's <u>just higher</u> than the <u>current</u> that'll <u>normally</u> flow in the circuit.

For example, if the <u>normal operating current</u> of the appliance is <u>4 A</u>, you'd pick a <u>5 A</u> fuse. This means that if the circuit is working normally, the fuse allows current to flow. But if the current gets <u>above 5 A</u>, the fuse will 'blow' and break the circuit, protecting the appliance from damage.

Use P = V × I to Work Out Current

1) Most electrical appliances show their <u>power rating</u> and <u>voltage rating</u>.

2) To pick the right <u>fuse</u>, you need to work out the <u>current</u> that the item will use. The formula you need is:

$$P = V \times I \quad \text{(see p.83)}$$

SUPPLY 240V — A / C 50HZ ONLY
MAX POWER CONSUMPTION: 45 WATTS.
DO NOT REMOVE ANY COVERS. DANGEROUS VOLTAGES EXIST INSIDE THIS UNIT
REFER ALL SERVICING TO QUALIFIED PERSONNEL.

3) But you're working out <u>current</u>, not power, so you'll need to rearrange the formula, using this handy <u>formula triangle</u>. ➡

EXAMPLE: A hairdrier is rated at 230 V, 1 kW. Find the fuse needed.

ANSWER: Cover up the I in the formula triangle — that gives you $I = \dfrac{P}{V}$

I = P/V = 1000/230 = 4.3 A.

Remember, the fuse should normally be rated just a little higher than the normal current, so a <u>5 amp fuse</u> is ideal for this one.

Circuit Breakers are Similar to Fuses but They're Resettable

<u>Circuit breakers</u> work in the same sort of way as a fuse — if a large current flows through them, they break the circuit.

1) A circuit breaker is a type of <u>electrical switch</u>. If the current flowing through it gets too high, the switch <u>opens</u> and cuts off the live supply.

2) Circuit breakers can be <u>reset</u> at the flick of a switch, so they're <u>more convenient</u> than fuses that have to be replaced each time they melt. Clever, eh?

Fuse rating — I'd say 9 out of 10 for usefulness...

If your fuse rating is <u>too low</u>, the fuse will blow every time you switch the appliance on — hardly convenient. If the fuse has <u>too high</u> a rating, it won't blow soon enough to protect your appliance.

Calculating Energy Usage and Cost

When you pay an electricity bill, it's the <u>energy</u> your appliances have used that you're paying for.

Power is Energy Used Per Second

More powerful appliances use energy more quickly. For example, take a <u>kettle</u>. The <u>higher the power</u> of the kettle, the <u>faster</u> it converts electrical energy into heat energy — and the <u>faster it boils</u>.

1) The units of energy are <u>joules</u>, J.
2) The <u>power</u> of an appliance tells you <u>how many joules</u> of electrical energy it converts into other forms of energy <u>per second</u>.
3) E.g. a 60 W light bulb converts 60 J of <u>electrical energy</u> into <u>light</u> energy and <u>heat energy</u> per second.
4) Here's the formula to work it out:

$$\text{POWER (in W)} = \frac{\text{ENERGY (in J)}}{\text{TIME (in s)}}$$

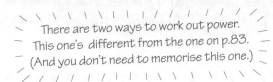
There are two ways to work out power. This one's different from the one on p.83. (And you don't need to memorise this one.)

5) It's important to get the right <u>units</u>. Make sure energy's in <u>joules</u> and time's in <u>seconds</u>.

EXAMPLE: A drill converts 24 kJ of <u>electrical energy</u> into <u>kinetic</u>, <u>heat</u> and <u>sound</u> energy in 1 minute. Calculate the power of the drill (in watts).

ANSWER: Energy needs to be in joules. 1 kJ = 1000 J, so 24 kJ = <u>24 000 J</u>.
Time needs to be in seconds. 1 minute = <u>60 s</u>.
Now use the formula. Power = 24 000 J ÷ 60 s = <u>400 W</u>.

Kilowatt-hours (kWh) are "UNITS" of Energy

1) Your electricity meter records how much <u>energy</u> you use in units of <u>kilowatt-hours</u>, or <u>kWh</u> (not J). A <u>kilowatt-hour</u> is the amount of electrical energy converted by a <u>1 kW</u> appliance left on for <u>1 hour</u>.
2) The energy an appliance uses depends on its <u>power</u> and the <u>time</u> it's on for.
3) And yes, there's a formula for it. (It's a rearrangement of the one above, but <u>with different units</u>.)

$$\text{ENERGY (in kWh)} = \text{POWER (in kW)} \times \text{TIME (in hours)}$$

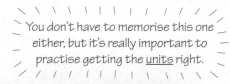
You don't have to memorise this one either, but it's really important to practise getting the <u>units</u> right.

4) If you know how much <u>energy</u> an appliance has used, it's fairly easy to calculate how much that energy <u>costs</u>:

$$\text{Total Cost} = \text{Number of kWh} \times \text{Cost per kWh}$$

EXAMPLE: An electricity supplier charges 12p per kWh. Find the cost of having an 8.5 kW shower on for 15 minutes.

ANSWER: Energy = Power × Time = 8.5 kW × 0.25 h = <u>2.125 kWh</u>.
Cost = Number of kWh × Cost per kWh = 2.125 × 12p = <u>25.5p</u>.

Convert energy quickly — eat a cream cake then run 10 miles...

You're <u>highly likely</u> to be asked about the <u>energy usage</u> of appliances, and the <u>cost</u> of the energy. So — How much energy does a 60 W light bulb use in 3 hours? See p.148 for answer.

Using Energy Efficiently

Electrical appliances convert electrical energy into other kinds of energy. Some of this energy is useful and some of it isn't.

Most Energy Transfers Involve Some Waste, Often as Heat...

1) <u>Useful devices</u> are only <u>useful</u> because they <u>convert energy</u> from <u>one form</u> to <u>another</u>.
2) In doing so, some of the useful <u>input energy</u> is always <u>'lost'</u> or <u>wasted</u>, often as <u>heat</u>.

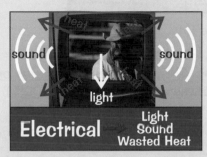

Electrical — Light Sound Wasted Heat

The useful energy produced by a <u>TV</u> is <u>light and sound energy</u>. But the TV also gets warm. (If you don't believe me, check for yourself.)

In a <u>kettle</u>, the useful energy is the <u>heat energy</u> of the <u>water</u>. But the outside of the kettle also gets hotter — and heats up the surrounding air.

Electrical — Useful Heat Wasted Heat

3) The <u>less energy</u> that's <u>wasted</u>, the <u>more efficient</u> the device is said to be.
4) The <u>energy flow diagram</u> is similar for <u>all devices</u>, really.
5) The important thing to remember is:

> <u>No</u> device is 100% efficient and some wasted energy is always <u>dissipated</u> as <u>heat</u>.

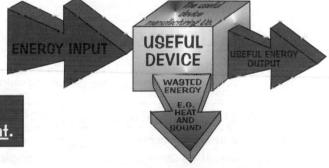

ENERGY INPUT — USEFUL DEVICE — USEFUL ENERGY OUTPUT — WASTED ENERGY E.G. HEAT AND SOUND

...But No Energy Disappears

1) Whenever energy is converted from one form to another, there's <u>always</u> some energy wasted — usually as <u>heat</u> and <u>sound</u>.
2) Take a power drill. Some of the electrical energy is converted into kinetic energy (the drill bit <u>moves</u> pretty quickly) but the drill also gets <u>warm</u> (heat energy) and makes a <u>noise</u> (sound energy).
3) <u>Heat</u> energy from the drill warms up the material that's being drilled into and the air around it — the energy is <u>spreading out</u> into the surroundings.
4) The <u>total</u> amount of <u>energy</u> is the <u>same</u> — it's still all there, but it <u>can't be easily used</u> or <u>collected back in</u> again. (That's why it's often called 'lost' energy. Aaahh.)

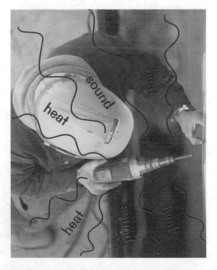

Let there be light — and a bit of wasted heat...

So, in a nutshell — useful output energy is always less than input energy, and heat's usually to blame. But remember, no energy has <u>gone</u> — it's just very hard to find. It's much the same as spilling a drink on the carpet — the drink's still there, just spread out all over the carpet and not much <u>use</u> any more.

Using Energy Efficiently

Energy's expensive, and most of us waste loads of it every day. But if your appliances are <u>efficient</u>, you won't waste too much of your money, and you'll be making a start on saving the planet. Good for you.

Efficiency is Often Given as a Percentage

1) Here's how to calculate the <u>percentage efficiency</u> of an electrical appliance:

$$\text{Efficiency} = \frac{\text{Useful Energy output}}{\text{Total Energy input}} \times 100\%$$

2) You find how much energy is <u>supplied</u> to a machine. (The total energy <u>INPUT</u>.)

3) You find how much <u>useful energy</u> the machine <u>delivers</u>. (The useful energy <u>OUTPUT</u>.)

4) Write down both numbers — <u>total input</u> and <u>useful output</u>. Then just <u>divide</u> the <u>smaller one</u> by the <u>bigger one</u> to get a value for <u>efficiency</u> somewhere between <u>0 and 1</u> (or <u>0 and 100%</u>). Easy.

<u>Example</u>: A toy car's motor uses 1000 J of energy to produce 300 J of kinetic (movement) energy. Find its efficiency.

<u>Answer</u>: Efficiency $= \dfrac{\text{useful energy output}}{\text{total energy input}} \times 100\% = \dfrac{300}{1000} \times 100\%$

$= 0.3 \times 100\% = \underline{30\%}$

Some Electrical Devices are More Efficient than Others

You nearly always have a <u>choice of devices</u> to use for any particular situation. For example, to light a room, you could use <u>filament lamps</u> ('ordinary' light bulbs) or you could use <u>low-energy</u> light bulbs.

Low-Energy Bulbs are More Efficient and They Last Longer...

For the same light output, a <u>low-energy</u> bulb is about <u>4 times as efficient</u> as an <u>ordinary</u> bulb. That means they use <u>one quarter</u> of the electrical energy that ordinary bulbs do. So:

1) Using low-energy bulbs could save you <u>money</u>.

2) If <u>everyone</u> used efficient bulbs, we'd need to generate less electricity in the first place — which could reduce damage to our environment.

Energy-efficient light bulbs also <u>last much longer</u>, so:

3) They're more <u>convenient</u> — you don't have to change them so often.

4) If <u>everyone</u> used them, we'd create <u>less waste</u> chucking old bulbs out, and we'd use up <u>less resources</u> and <u>less energy</u> making new ones.

... But They're Not Always the Best Choice

1) Low-energy bulbs are <u>more expensive</u> to buy.

2) Some <u>light fittings</u> don't take low-energy bulbs.

3) <u>Energy efficient</u> bulbs take a few <u>minutes</u> to get to <u>full brightness</u>, which can be inconvenient.

Don't waste your energy — turn the TV off while you revise...

And for 10 bonus points, calculate the efficiency of these machines:
TV — input energy 200 J, output light energy 5 J, output sound energy 2 J, output heat energy 213 J.
Loudspeaker — input energy 35 J, output sound energy 0.5 J, output heat energy 34.5 J. Answers p.148.

Heat Transfer

Heat energy tends to <u>flow away</u> from a hotter object to its <u>cooler surroundings</u>.
But then you probably knew that already. You also need to know <u>how</u> heat flows, so read on...

Heat is Transferred in Three Different Ways

1) Heat energy can be transferred by <u>radiation</u>, <u>conduction</u> or <u>convection</u>.

2) <u>Thermal radiation</u> (see below) is the transfer of heat energy by <u>electromagnetic waves</u> (see p.103)

3) <u>Conduction and convection</u> (see p.90) involve the transfer of energy by <u>particles</u>. <u>Conduction</u> is the main form of heat transfer in <u>solids</u>, and <u>convection</u> is the main form of heat transfer in <u>liquids and gases</u>.

4) <u>Thermal radiation</u> happens in <u>solids, liquids and gases</u>. Any object can absorb or emit heat radiation.

5) The <u>bigger the temperature difference</u>, the <u>faster heat is transferred</u> between an object and its surroundings. Kinda makes sense.

A hot dog absorbing some infrared radiation. Mmm... hot dog...

Thermal Radiation Transfers Heat by Waves

<u>Thermal radiation</u> can also be called <u>infrared radiation</u>, and it consists purely of electromagnetic waves.

1) <u>All objects</u> are <u>continually</u> emitting and absorbing <u>heat radiation</u>.

2) An object that's <u>hotter</u> than its surroundings <u>emits more radiation</u> than it <u>absorbs</u> (and <u>cools</u> down). An object that's <u>cooler</u> than its surroundings <u>absorbs more radiation</u> than it <u>emits</u> (and <u>warms</u> up).

3) You can <u>feel</u> this <u>heat radiation</u> if you stand near something <u>hot</u> like a fire or if you put your hand just above the bonnet of a recently parked car.

Recently parked car

After an hour or so...

Thermal Radiation is Increased By:

LARGE SURFACE AREA

1) <u>Heat</u> is <u>radiated</u> from the <u>surface</u> of an object. The <u>bigger</u> the <u>surface area</u>, the <u>more waves</u> can be <u>emitted</u> from the surface — so the <u>quicker</u> the <u>heat transfer</u>.

2) This is why <u>car and motorbike engines</u> often have '<u>fins</u>' — they <u>increase</u> the <u>surface area</u> so heat is radiated away quicker. So the <u>engine cools faster</u>.

3) It's the same with <u>heating</u> something up — the bigger the surface area exposed to the heat radiation, the <u>quicker it'll heat up</u>.

Cooling fins on engines increase surface area to speed up cooling.

DARK COLOUR AND ROUGH TEXTURE OF SURFACE

1) <u>Dark matt</u> surfaces <u>absorb</u> heat radiation falling on them much <u>better</u> than <u>light glossy</u> surfaces. They also <u>emit much more</u> heat radiation (at any given temperature).

2) <u>Silvered</u> surfaces (like tinfoil) <u>reflect</u> most of the heat radiation falling on them.

Thermal radiation — how to dry your winter underwear...

<u>Thermal radiation</u> is how we get heat from the <u>Sun</u> — it's the only way that heat's transferred in space. Remember, <u>everything</u> emits and absorbs radiation. How <u>quickly</u> this happens depends on the <u>temperature</u> of the object, the <u>colour</u> and <u>texture</u> of its <u>surface</u> and its <u>surface area</u>. All worth learning.

Heat Transfer

Thermal radiation is only one method of heat transfer. The others are <u>conduction</u> and <u>convection</u>.

Conduction of Heat Occurs Mainly in Solids

1) <u>Conduction of heat</u> is where <u>vibrating particles</u> pass on their <u>extra energy</u> to <u>neighbouring particles</u>.

2) This process continues <u>throughout the solid</u> and gradually some of the <u>extra energy</u> (or <u>heat</u>) is passed all the way through the solid.

3) This causes a <u>rise in temperature</u> at the other side of the solid, and so heat radiates from its surface more quickly.

4) <u>Metals conduct</u> heat <u>better</u> than <u>plastic or wood</u>, which is why <u>pans</u> are made from <u>metal</u>, but their <u>handles</u> are made of <u>plastic or wood</u>.

PAN HANDLE EXAMPLE

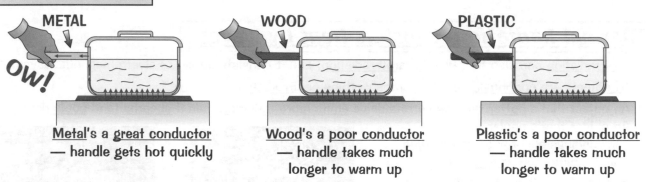

Metal's a <u>great conductor</u> — handle gets hot quickly

Wood's a <u>poor conductor</u> — handle takes much longer to warm up

Plastic's a <u>poor conductor</u> — handle takes much longer to warm up

Convection of Heat — Liquids and Gases Only

1) <u>Gases and liquids</u> are usually free to <u>slosh about</u> — and that allows them to transfer heat by <u>convection</u>, which is a <u>much more effective process</u> than conduction.

2) <u>Convection</u> occurs when the more energetic particles <u>move</u> from the <u>hotter region</u> to the <u>cooler region</u> — <u>and take their heat energy with them</u>. (Note: convection simply <u>can't happen in solids</u> because the particles <u>can't move</u>.)

3) This is how <u>immersion heaters</u> in <u>kettles</u> and <u>hot water tanks</u> work.

IMMERSION HEATER EXAMPLE

1) <u>Heat energy</u> is <u>transferred</u> from the heater coils to the water by conduction (particle collisions).

2) The <u>particles</u> near the coils get <u>more energy</u>, so they start <u>moving faster</u>. This makes the water <u>expand</u> and become <u>less dense</u> — so it rises.

3) As the <u>hot water</u> rises, the <u>colder</u> water at the top of the tank <u>sinks</u> to the bottom...

4) ...where it's <u>heated by the coils</u> and rises — and so it goes on. You end up with <u>convection currents</u> going up, round and down, <u>circulating</u> the heat through the water.

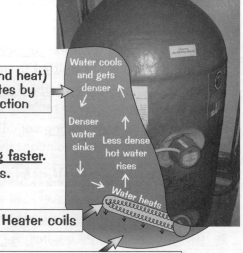

Water (and heat) circulates by convection

Water cools and gets denser

Denser water sinks

Less dense hot water rises

Water heats

Heater coils

Water stays cold below the heater

Ah well — only a couple of hours till teatime...

Well OK, this isn't the best fun you've ever had, but it's not exactly rocket science either — it's all fairly straightforward, common-sense stuff. Think of a few <u>real</u> examples and it'll all make a lot more sense.

Heat Transfer in the Home

In the UK it can be expensive to keep a house <u>warm</u> in winter — because heat <u>flows</u> outside into the cold.

Effectiveness and Cost-Effectiveness are Not the Same...

To reduce unwanted <u>heat transfer</u> from a house, you need <u>insulation</u>. There are several types:

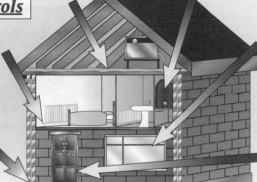

Loft Insulation
Initial Cost: £200
Annual Saving: £50
Payback time: <u>4 years</u>

Hot Water Tank Jacket
Initial Cost: £15
Annual Saving: £15
Payback time: <u>1 year</u>

Thermostatic Controls
Initial Cost: £100
Annual Saving: £20
Payback time: <u>5 years</u>

Double Glazing
Initial Cost: £3000
Annual Saving: £60
Payback time: <u>50 years</u>

Cavity Wall Insulation
Initial Cost: £500
Annual Saving: £70
Payback time: <u>7 years</u>

Draught-Proofing
Initial Cost: £50
Annual Saving: £50
Payback time: <u>1 year</u>

1) It <u>costs money</u> to buy and install insulation, but it will <u>reduce your heating bills</u>.

2) Eventually, the <u>money you've saved</u> on heating bills will <u>equal</u> the <u>initial cost</u> of installing the insulation — the time this takes is called the <u>payback time</u>. (The figures given above are just a rough guide.)

3) <u>Cheaper</u> methods of insulation are usually <u>less effective</u> — they tend to save you less money per year — but they often have <u>shorter payback times</u>.

4) If you <u>subtract</u> the <u>annual saving</u> from the <u>initial cost</u> repeatedly then <u>eventually</u> the one with the <u>biggest annual saving</u> must always come out as the winner, if you think about it.

5) But you might sell the house (or die) before that happens. If you look at it over, say, a <u>five-year period</u> then a cheap and cheerful <u>hot water tank jacket</u> wins over expensive <u>double glazing</u>.

Know Which Types of Heat Transfer Are Involved:

1) <u>CAVITY WALL INSULATION</u> — foam in the gap between the bricks can reduce <u>convection</u> and <u>radiation</u> across the gap.

2) <u>LOFT INSULATION</u> — a thick layer of fibreglass wool laid out across the whole loft floor reduces <u>conduction</u>.

3) <u>DRAUGHT-PROOFING</u> — strips of foam and plastic around doors and windows stop draughts of cold air blowing in, i.e. they reduce heat loss due to <u>convection</u>.

4) <u>DOUBLE GLAZING</u> — two layers of glass means more <u>radiation</u> reflected back, and the air gap between the layers reduces <u>conduction</u>.

5) <u>THERMOSTATIC RADIATOR VALVES</u> — these simply prevent the house being <u>over-warmed</u>.

6) <u>HOT WATER TANK JACKET</u> — insulating material (e.g. fibreglass wool) reduces <u>conduction</u>.

7) <u>THICK CURTAINS</u> — big bits of cloth over the window reduce heat loss by <u>convection</u> and <u>radiation</u>.

It's payback time...

And it's the same with, say, cars. Buying a very fuel-efficient car might sound like a great idea — but if it costs loads more than a clapped-out old fuel-guzzler, you might still end up out of pocket. To work out <u>cost-effectiveness</u>, you always have to offset the initial cost against the estimated annual savings.

Useful Mixtures in the Home

Chemists make loads of things that we use round the home — and almost all of them are mixtures of some kind, rather than pure substances. A solution is one such useful mixture...

A Solution is a Mixture of Solvent and Solute

1) A solution is a substance (the solute) dispersed in a liquid (the solvent) to form a see-through liquid.

2) Some everyday examples are instant coffee, soluble aspirin and bath salts.

3) A solution is different from other mixtures (see next page) because the bonds holding the solute molecules together actually break, and the molecules mix completely with the molecules in the solvent — this is called dissolving.

4) Water is a very common solvent. In fact, the water you get from the tap already has lots of substances dissolved in it.

5) A solution can be separated by heating it to evaporate off the liquid, leaving just the solid. This is one way to get salt from the sea — in warm countries, the Sun's heat is used to evaporate the water, leaving salt behind. →

Many Substances are Insoluble in Water...

1) Some paints don't dissolve in water, but are soluble in some organic solvents.

2) For example, oil paint can be dissolved in turpentine, which is an organic solvent. So turpentine can be used as a paint remover or thinner for oil paint.

3) Likewise, ethanol is used in cosmetics and smelly toiletries, because it can dissolve the nice-smelling perfume oils — and can be mixed with water.

4) Loads of different substances can be used as solvents — but you have to use the right solvent for the right situation.

5) The ability to dissolve a solute isn't the only consideration though... some solvents are very toxic or give off nasty fumes, so there may be situations where you really wouldn't want to use them.

Gases are Also Soluble — This is How Fizzy Drinks Work

1) All gases are soluble to some extent. For example, loads of fish survive on oxygen dissolved in oceans, swimming pools contain dissolved chlorine gas, and the 'fizz' in fizzy drinks comes from dissolved carbon dioxide.

2) The amount of gas you can dissolve depends on the temperature and pressure — at high pressures and low temperatures, more gas will dissolve.

3) That's why fizzy drinks are bottled at high pressures, so that they can dissolve lots of carbon dioxide. Then when you take the lid off, the pressure decreases and the liquid can't hold as much carbon dioxide in solution — so it comes out in bubbles.

Learn this page — it's the only solution...

If you ever spill bright pink nail varnish (or any other colour for that matter) on your carpet, go easy with the nail varnish remover. If you use too much, the nail varnish dissolves in the remover, forming a solution which can go everywhere — and you can end up with an enormous bright pink stain... aaagh.

Useful Mixtures in the Home

As well as the more familiar and straightforward <u>solution</u> (p.92), we use a few other liquidy-type mixtures around the home that you need to know about: <u>suspensions</u>, <u>emulsions</u>, <u>gels</u>, <u>foams</u> and <u>aerosols</u>.

A Suspension is a Cloudy Mixture of a Solid in a Liquid

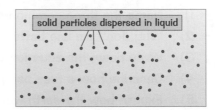

solid particles dispersed in liquid

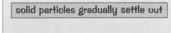

solid particles gradually settle out

1) A suspension is a mixture of <u>small solid particles</u> (e.g. silt) mixed in with a liquid (e.g. water).

2) But in a suspension, the particles aren't small enough to stay floating around — eventually they will <u>separate</u>, as the solid particles gradually <u>settle</u> to the bottom.

3) <u>Muddy water</u> is a great example of a suspension — get a jar of muddy water, leave it on the side for an hour or two, and you'll end up with <u>clear water</u> with a <u>layer of sludge</u> at the bottom. Nice.

A Gel is a Jelly-Like Mixture of a Solid in a Liquid

1) A <u>gel</u> is made from <u>tiny solid particles</u> dispersed in liquid, in a way that the solid particles 'trap' the liquid. This means the whole thing behaves a bit <u>more like a solid</u> than a <u>liquid</u>.

2) <u>Jelly</u>, <u>hair gel</u> and some <u>toothpastes</u> are gels.

3) Some gels can <u>set</u> and become <u>hard</u>, which is why they're useful for things like <u>hair styling</u> — slap on a load of <u>hair gel</u>, sculpt it into a <u>nice quiff</u> and leave to set. Grand.

Hair gel — mmm... gloopy

(just don't get too carried away)

An Emulsion is a Cloudy Mixture of a Liquid in Another Liquid

1) An <u>emulsion</u> is made from <u>tiny droplets of one liquid dispersed in another</u>, where the liquids won't properly dissolve in each other, e.g. oil and water.

2) <u>Emulsion paint</u>, <u>mayonnaise</u> and some <u>salad dressings</u> are examples of emulsions.

3) Emulsions are <u>cloudy</u>, and will <u>eventually separate</u> if left for a while. E.g. you can make salad dressing from <u>olive oil</u> and <u>vinegar</u>. You <u>shake it up</u> to make a <u>cloudy liquid</u> then drench your lettuce with it. But if you come back to it <u>tomorrow</u> lunchtime, the oil will have <u>separated out</u> and be sitting on top of the vinegar, so you'll have to <u>shake it up again</u>.

With Foams and Aerosols There's Gas Involved

1) A <u>foam</u> is made of <u>tiny gas bubbles</u> dispersed in either a solid, e.g. bread, or a liquid, e.g. <u>shaving foam</u> or the head on <u>beer</u>.

2) An <u>aerosol</u>, on the other hand, is <u>tiny droplets of a liquid</u> dispersed in a gas, e.g. <u>hairspray</u> or <u>spray paint</u>.

Foaming at the mouth? — Don't bother learning this, it's too late...

These mixtures have <u>so many uses</u> round the home that just giving them a page seems a bit unfair. But this is all you need to know, so I'll not bother your already crammed brain with any more details. For each mixture, learn the <u>structure</u>, at least <u>one use</u>, and say <u>why it's suited</u> to its use.

Revision Summary for Section 2.8

Science at school, science at home — you really can't get away from it. There's no getting away from exams either, so here are some practice questions — and if you can't do them all, you know what to do.

1) Name the three common energy sources used for heating homes.

2) Why don't we use batteries to provide electricity for cookers and washing machines?

3) Write down the equation which links power, voltage and current.

4) Describe how a fuse can protect an electrical device from damage.

5) Explain how the earth wire and fuse in a plug can protect you from electric shocks.

6)* Bill has a circular saw. It's rated 230 V, 1.5 kW. What fuse should it have — 5 A, 7 A or 13 A?

7) Describe how a circuit breaker works.

8) In what way is it better to have a circuit breaker than a fuse?

9) What are the units of: a) energy, b) power?

10)* Bill uses his 550 W power drill for half a minute. How much energy has been transferred?

11)* What is a kWh a unit of? Change 2 kWh into joules.

12)* How many kWh will you use if you leave a 100 W light bulb on for: a) 3 hours, b) 30 minutes?

13)* Freda's electricity supplier charges 11 p per kWh.
How much will it cost her to have her 200 W hair straighteners on for 15 minutes?

14) Describe the energy transfers that take place in a light bulb.

15)* A light bulb produces 40 J of light energy for every 100 J of electrical energy it uses.
a) What is its percentage efficiency?
b) How much energy does the light bulb waste for every 100 J of energy it uses?

16)* A different light bulb is advertised like this: "Only wastes 230 J for every 1000 J of energy input."
Calculate the percentage efficiency of this light bulb.

17) More efficient light bulbs aren't always a better choice. Why not?

18) What is the main method of heat transfer in: a) solids, b) liquids, c) gases?

19) What happens to the heat that's 'lost' when a cup of hot tea cools down?

20) Which will lose heat fastest — a hot water bottle filled with water at 90 °C or an identical hot water bottle with water at 81 °C? (Both are left in the same conditions.)

21) How is heat radiation from a surface affected by: a) its colour, b) its texture, c) its surface area?

22) Explain why saucepans are made of metal but saucepan handles are made of wood or plastic.

23) What happens to the particles in a metal saucepan when it's heated?

24) Why does air rise when it's heated?

25) In an immersion heater, how is heat transferred:
a) from the heater coils to the water?
b) from the bottom of the tank to the top of the tank?

26) How could you reduce heat loss from a house through: a) the roof, b) the windows, c) the walls?

27) What is payback time?

28) What is: a) a solution, b) a suspension, c) a gel, d) an emulsion?

29) How can you separate the components of a solution?

30) Why are fizzy drinks fizzy?

31) How can you separate out the solid from a suspension?

32) Why do the instructions on emulsion paint tell you to 'stir well' before using it?

33) What's the difference between a foam and an aerosol?

*See p.148 for answers.

Speed

There are thousands of road accidents in the UK every year, and many of them are caused by people <u>driving too fast</u>. That's why there are often <u>speed cameras</u> at accident 'blackspots' — to try and deter people from breaking the legal speed limit.

Speed is Distance Travelled Divided by Time

1) To find out the <u>speed</u> of an object, you need to know how <u>far</u> it travels in a certain <u>time</u>.

2) When the distance is measured in <u>metres</u> (m) and the time in <u>seconds</u> (s), the speed is calculated in <u>metres per second</u> (m/s).

3) You need to <u>learn</u> this formula and practise using it:

$$\text{Speed (in m/s)} = \frac{\text{Distance (in m)}}{\text{Time (in s)}}$$

Watch out for units — if you're given distance in <u>km</u> or time in <u>minutes</u>, say, you'll need to change them into <u>m</u> and <u>s</u> first.

<u>EXAMPLE</u>:

A speed camera took two photos of a car. The photos were taken <u>0.5 seconds</u> apart and the car travelled <u>6 metres</u> in this time. Find the speed of the car.

<u>ANSWER</u>: speed = $\dfrac{\text{distance travelled}}{\text{time taken}}$ = $\dfrac{6 \text{ m}}{0.5 \text{ s}}$ = <u>12 m/s</u>.

4) An object's <u>velocity</u> is a bit different from its speed. <u>Velocity</u> means speed <u>in a particular direction</u>. Two objects could be going at the same speed, but have different velocities.

E.g. if Geoff and Dave are driving in the <u>same direction</u>, <u>both going at 70 mph</u>, they have the <u>same velocity</u>. But if they're travelling at 70 mph in <u>opposite directions</u>, they have <u>different velocities</u>.

Fortunately, you don't need to worry about this too much — just don't be put off if exam questions use 'velocity' instead of 'speed'.

You Might Have to Rearrange the Formula

You might have to find the <u>distance travelled</u> or the <u>time taken</u> rather than the speed. You'd need to rearrange the formula or use a nice formula triangle.

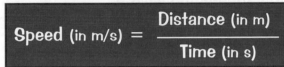

<u>EXAMPLE</u>: A bus is moving at 12 m/s. A man runs at 7 m/s to try and catch it.
a) How long will it take the bus to move 5 m?
b) How far will the man move in this time?

<u>ANSWER</u>: a) You need to find <u>time</u>, so cover up the 't' in the formula triangle and then:
Time taken = d ÷ s = 5 ÷ 12 = <u>0.417 s</u> (to 3 d.p.).

b) It's <u>distance</u> you're after here:
Distance travelled = s × t = 7 × 0.417 s = <u>2.92 m</u>.

Don't speed through this page — learn it properly...

<u>Speed cameras</u> measure the speed of motorists using two photos, taken a fraction of a second apart. Lines painted on the road help them work out <u>how far</u> the car travelled between the photos — and so <u>how fast</u> it was going. And the photos always have the car's number plate in them. Clever, eh?

Acceleration

Some people are very fond of their <u>cars</u>. And some people are very proud of their car's <u>acceleration</u> — they say things like, "Eeeh, Geoff, I managed 0 to 60 in 7 seconds in this'un t'other day." (Well, that's if they have friends called Geoff and they're from Yorkshire.) Anyway, acceleration's important — read on.

Acceleration is How Quickly You're Speeding Up

Acceleration is definitely <u>not</u> the same as <u>speed</u>. Acceleration is how <u>quickly</u> the speed is <u>changing</u>.

Here's the lovely formula for acceleration (and yes, you need to learn this one too):

$$\text{Acceleration (in m/s}^2) = \frac{\text{Change in speed (in m/s)}}{\text{Time (in s)}}$$

Strictly speaking it's the change in <u>velocity</u> (see p.95), but you don't need to worry about that.

There are a couple of <u>slightly tricky things</u> to watch out for with this formula:

1) You've got to work out the '<u>change in speed</u>', as shown in the example below, rather than just putting in a single value for speed.

2) Secondly there's the <u>units</u> of acceleration, which are <u>m/s²</u>. <u>Not m/s</u>, which is <u>speed</u>, but <u>m/s²</u>. Got it? Splendid. But just in case: <u>Not m/s</u>, but <u>m/s²</u>.

<u>EXAMPLE</u>:

Geoff's car accelerates steadily from <u>10 m/s</u> to <u>40 m/s</u> in <u>12 seconds</u>.
Calculate its acceleration.

> <u>ANSWER:</u>   First calculate the <u>change in speed</u> (it's pretty easy here): 40 − 10 = 30 m/s.
> You're given the time taken for the change (12 s), so there's nothing to work out there.
>
> So, using the formula, acceleration $= \dfrac{\text{change in speed}}{\text{time}} = \dfrac{30 \text{ m/s}}{12 \text{ s}} = \underline{2.5 \text{ m/s}^2}$.

Deceleration is How Quickly You're Slowing Down

1) In vehicles, <u>decelerating</u> (slowing down) can be much more dangerous than accelerating — because you're more likely to have to slow down <u>suddenly</u> (to miss a little puppy dog in the road, say).

2) To <u>calculate</u> deceleration, you do exactly the same as you do for acceleration.

<u>EXAMPLE</u>:

Geoff takes 10 seconds to slow down from 40 m/s to 28 m/s.
Calculate his deceleration.

<u>ANSWER</u>:

Again, find the change in speed first: 40 − 28 = 12 m/s.
The time's already given (10 s) so:
Deceleration = 12 ÷ 10 = <u>1.2 m/s²</u>.

Deceleration — picking the celery out of your salad...

Where would we be without acceleration? In bed, that's where — because if you start off stationary (i.e. with <u>zero velocity</u> — <u>asleep in bed</u>) you have to <u>accelerate</u> to get moving at all. Darn it.

Cars and Stopping Distances

If you need to <u>stop</u> quickly on a bicycle, in a car, lorry or whatever, you put the <u>brakes</u> on (obviously). So far so good. But if your brakes are worn or the road's wet and slippy or you were going too fast then you won't be able to slow down in time to avoid the kitten / puppy / another cute animal of your choice.

Many Factors Affect Your Total Stopping Distance

1) The <u>stopping distance</u> of a car is the distance covered in the time between the driver <u>first spotting</u> a hazard and the car coming to a <u>complete stop</u>.

2) The stopping distance is the <u>thinking distance</u> plus the <u>braking distance</u>:

1) Thinking Distance

This is the distance the car travels in the time between the driver <u>noticing the hazard</u> and <u>applying the brakes</u>. It's affected by <u>two main factors</u>:

a) The driver's reaction time
The time it takes the driver to apply the brakes once they've noticed the hazard. This is generally longer if the driver is <u>old</u>, <u>tired</u>, has drunk <u>alcohol</u> or taken certain other <u>drugs</u>.

b) The car's speed
Whatever your reaction time, the <u>faster</u> you're going, the <u>further</u> you'll go during that time — remember that d = s × t.

2) Braking Distance

This is the distance the car travels during its deceleration while the brakes are being applied. It's affected by <u>four main factors</u>:

a) The car's speed

The <u>faster</u> the car was going, the <u>further</u> it will take to stop.

b) How heavily loaded the car is

<u>Heavily laden</u> vehicles take <u>longer to stop</u> (for the same braking force). A car won't stop as quickly when it's full of people and luggage and towing a caravan.

c) How good the brakes are

<u>Worn</u> or <u>faulty</u> brakes will be less effective at slowing the car down.

d) How good the grip on the road is

This depends on the <u>road surface</u>, the <u>weather</u> and the condition of the car's <u>tyres</u>. If the road is really smooth, or it's wet or icy, or your tyres are badly worn, your braking distance will be much greater.

<u>Bad visibility</u> can also be a major factor in accidents — lashing rain, thick fog, bright oncoming lights, etc. might mean that a driver <u>doesn't notice</u> a hazard until they're quite close to it — so they have a much shorter distance available to stop in.

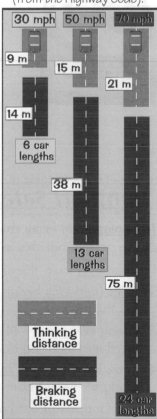

Typical stopping distances (from the Highway Code).

30 mph · 50 mph · 70 mph

9 m · 15 m · 21 m

14 m

6 car lengths

38 m

13 car lengths

75 m

Thinking distance

Braking distance

24 car lengths

Stop right there — and learn this page...

Drivers should <u>never get too close</u> to the vehicle in front — it's important to leave enough distance to stop in an emergency. Motorway pile-ups happen when people drive too fast and too close together. It's wise to <u>slow down</u> in heavy <u>rain</u> and when the roads are likely to be <u>icy</u>. It's also wise to <u>check your tyres</u> regularly — in fact it's illegal to drive with tyres that are too worn (less than 1.6 mm tread depth).

Transport Safety

There are <u>rules and regulations</u> to make sure <u>drivers</u> don't put <u>themselves</u> and <u>others</u> at <u>unnecessary risk</u>.

Transport Laws are There to Protect the Driver and the Public

1) <u>There are Rules</u> The Highway Code states <u>how</u> drivers, cyclists, pedestrians and horse-riders <u>should behave</u> on public roads.

2) <u>Drivers Have to Pass Tests</u> You have to pass a <u>theory test</u>, a <u>hazard perception test</u> and a <u>practical test</u> before you're allowed to drive a car, motorbike etc. This is all to make sure you <u>know the rules</u> and can <u>drive safely</u>.

3) <u>Drink-Driving is Illegal</u> It's <u>illegal</u> to <u>drink and drive</u>, because alcohol affects a driver's <u>reaction time</u>, <u>judgement</u> and <u>coordination</u>. The police can use a <u>breathalyser test</u> to check a driver's alcohol level.

4) <u>The MOT Test Checks that Cars are Roadworthy</u>

All cars over 3 years old must have an **MOT** test once a year. This is to check that the car is '<u>roadworthy</u>' — isn't about to fall apart. It also checks the car's exhaust gases etc. against <u>environmental regulations</u>.

5) <u>There are Signs Giving Information</u>

<u>Driving too fast</u> is very <u>dangerous</u> so there are legal <u>speed limits</u>. The limits vary — they're highest on motorways and dual carriageways, and lowest on roads in towns and villages. Other road signs give <u>information</u>, e.g. warning about hazards ahead — sharp bends, sheep likely to be crossing, etc.

6) <u>You Can't Use a Mobile Phone Unless it's Hands Free</u>

Using a mobile handset takes the driver's <u>hands off the wheel</u> and can be <u>distracting</u>. Drivers are <u>four times more likely to have an accident</u> when using a mobile phone than when they're fully concentrating.

Transport Safety Can be Improved by Technology...

<u>Technology</u> can make road transport safer for <u>drivers</u> and <u>passengers</u>. You don't have to remember all the details, but this is the kind of thing:

1) <u>Seatbelts</u> keep the driver and passengers safer if the car stops suddenly.

2) <u>ABS</u> (<u>anti-lock braking system</u>) can <u>stop the car skidding</u> during hard braking, so the driver can <u>steer</u> as they stop.

3) Many vehicles are fitted with <u>airbags</u> — to stop people hitting hard surfaces and slow them down more gently in a crash.

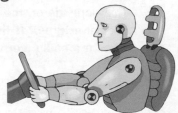

4) Cars have <u>crumple zones</u> designed into them — in a collision, the metal body of the car crumples up, reducing some of the force of the impact.

5) New models of car have to be thoroughly tested for safety. Manufacturers use <u>crash test dummies</u> to do this, which is nice.

So does the Tardis have an MOT for every new series?

Modern cars may have more safety features than, say, a Model T Ford, but they're also generally bigger and they go faster. And, sadly, all the safety tests, seatbelts and airbags in the world are no help to the poor pedestrian or cyclist who gets hit at 60 mph by a <u>careless driver</u> in a great big 4×4.

Fuel for Transport

Most transport fuels (like petrol and diesel) come from crude oil, which is dug up from deep underground.

Petrol and Diesel Come from Crude Oil

1) Crude oil is a fossil fuel that's made up of a mixture of hydrocarbons (p.56).

2) Hydrocarbons are compounds. The molecules in hydrocarbons contain only hydrogen and carbon atoms.

3) Petrol and diesel are both hydrocarbons.

Petrol contains shortish hydrocarbons (about 8 carbon atoms in each molecule)

PETROL — e.g. C_8H_{18}

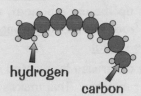

hydrogen

carbon

Diesel has longer molecules with about 12 carbon atoms.

DIESEL — e.g. $C_{12}H_{26}$

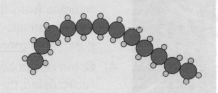

Hydrocarbons Release Energy When They Burn

1) When hydrocarbons burn in plenty of oxygen, energy is released as heat. This is the word equation for the reaction:

This reaction is called complete combustion.

$$\text{hydrocarbon} + \text{oxygen} \rightarrow \text{carbon dioxide} + \text{water} (+ \text{heat})$$

2) You might have to write balanced symbol equations for this (see p.68). For example:

$$C_6H_{12} + 8\,O_2 \rightarrow 5\,CO_2 + 6\,H_2O\ (+ \text{energy})$$

The energy is the useful bit.

3) The number of oxygen atoms needed depends on the size of the hydrocarbon molecule.

4) The bigger the molecule, the more carbon and hydrogen there is...

5) ...and so the more oxygen you need for combustion...

6) ...and the more carbon dioxide and water you end up with in total after the reaction. There's a pattern:

$$C_6H_{14} + 9\tfrac{1}{2}\,O_2 \rightarrow 6\,CO_2 + 7\,H_2O\ (+ \text{energy})$$
$$C_7H_{16} + 11\,O_2 \rightarrow 7\,CO_2 + 8\,H_2O\ (+ \text{energy})$$
$$C_8H_{18} + 12\tfrac{1}{2}\,O_2 \rightarrow 8\,CO_2 + 9\,H_2O\ (+ \text{energy})$$
$$C_9H_{20} + 14\,O_2 \rightarrow 9\,CO_2 + 10\,H_2O\ (+ \text{energy})\ ...\text{etc.}$$

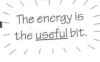 *petrol*

You could use this pattern to work out the equations for other hydrocarbons, such as $C_{10}H_{22}$. In fact, it would be very wise to make sure you can before you go into the exam.

Crude oil — pollutes the air with its dirty jokes...

The good thing about hydrocarbons is that the clue's in the name — they're made of hydrogen and carbon — and nothing else. You'll need to practise balancing those combustion equations though.

Fuel Efficiency and Energy Transfer

Petrol and diesel are burned in vehicle engines to release heat energy. An engine converts (transfers) heat energy into kinetic energy. The more efficient this energy transfer is, the less fuel is needed.

Efficiency Means the Percentage of Energy That's Useful

Energy transfer just means converting energy from one form into another. Whenever energy is transferred some of it is wasted, usually as heat (see p.87). When you're burning fuels, it's generally heat you're after — but only if it does something useful. Take a car engine for example:

1) When the fuel is burned its chemical energy is changed into heat energy.

2) The engine turns this heat energy into kinetic energy and makes the car move. The engine also charges the battery (which powers the lights, windscreen wipers, etc.) and operates systems like power steering.

3) Some of the heat energy released from the fuel is wasted. E.g. heat radiates from the engine and warms up the car's bonnet, which then warms the air. None of the energy from the fuel disappears — it just gets much more spread out and less useful.

4) Some of the heat energy that might otherwise be wasted can be recirculated through the car's heater system. This makes the car a bit more efficient as a whole because more of the energy output is used. (Generally, though, when people talk about how 'fuel efficient' a car is, they mean how far it goes per litre of fuel, not how cosy it is.)

Incomplete Combustion is Inefficient...

1) Complete combustion only happens if the fuel is burned where there is plenty of oxygen (see p.56).

2) If there's not enough oxygen, combustion will be incomplete.

3) In incomplete combustion, the products include carbon (C) and carbon monoxide (CO) as well as carbon dioxide and water. E.g.

$$C_5H_{12} + 5O_2 \rightarrow CO_2 + 2CO + 2C + 6H_2O \text{ (+ energy)}$$

Compare this with the equation for complete combustion of C_5H_{12} on p.99.

4) There's usually some incomplete combustion in most vehicle engines.

5) Incomplete combustion releases less energy than complete combustion. That means there's less energy available to drive the engine (as well as more pollution emitted — see below). So it's sensible to have engines serviced regularly to make sure they are running as efficiently and cleanly as possible.

...and Dangerous

Incomplete combustion isn't just inefficient — it releases dangerous pollutants into the air.

1) Carbon monoxide is a highly poisonous gas. If you inhale carbon monoxide, it joins with the haemoglobin in your blood (see p.13) and prevents it carrying oxygen around your body. This can be lethal — if you don't get enough oxygen to your cells you'll die.

2) The carbon that's produced can also be harmful. It forms very small particles called soot, which float around in the air as smoke. Inhaling this smoke can cause breathing problems or lung disease.

How an engine works — suck, squeeze, bang and blow...

Combustion is jolly useful — we'd be pretty cold without all the energy it releases (and there'd be no hot meals or going for a spin in the MG). Incomplete combustion is bad news, though. Sitting in a car in a traffic jam, especially in a built-up area, you breathe in a fair amount of muck from vehicle exhausts.

Alternative Fuels

Burning fossil fuels in cars, trains, planes etc. has a big impact on our environment (see p.63). But without fuels, we can't go anywhere or move stuff about. People don't like sitting at home all day with no stuff, so scientists are developing alternatives to fossil fuels.

Energy Content is Important...

Petrol and diesel make good fuels because they have a high energy content.

1) Energy content means how much energy is released when 1 kg of fuel OR 1 litre of fuel is burnt. You might be given an energy content graph (or some data) in the exam.

2) It's important to read energy content graphs very carefully — check the units and the scale. The units here are megajoules per kilogram (1 MJ = 1 000 000 J).

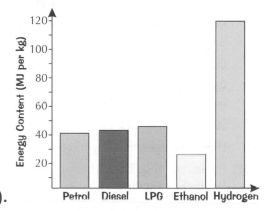

3) In theory, the higher the energy content, the further you could drive on one kilogram of fuel. So it looks like hydrogen's best... but it's not quite that simple.

4) If you made a similar graph in megajoules per litre, it would look different. E.g. 1 litre of LPG (liquefied petroleum gas) contains much less energy than a litre of petrol or diesel. It's the same for hydrogen (which is a gas).

5) Fuels with a low energy content per litre are inconvenient to use because they take up a lot of room in your car or bus, etc. You couldn't store so many joules of energy — so you'd have to keep filling up.

...but It's Not the Only Thing to Consider

There are other important things to think about:
Is the fuel safe?   Does it cause a lot of pollution?   How expensive is it?   Is there a reliable supply?
For example, hydrogen's pretty explosive and expensive and not many filling stations sell it.

Example — Gasohol from Sugar Cane in Brazil

Gasohol is a mixture of petrol and ethanol. It's used widely in Brazil.

1) Ethanol can be produced from plants like sugar cane — and Brazil grows lots of sugar cane. So there's a reliable and fairly cheap supply.

2) Ethanol is a liquid — so it's fairly easy to transport in tanks. And it's not particularly dangerous — unless you drink it.

His car probably runs on gasohol.

3) Its energy content is reasonable (though lower than petrol or diesel).

4) The only products when ethanol burns are water (harmless) and carbon dioxide. This CO_2 was taken in by the sugar cane as it grew, so it doesn't add to climate change. Air quality in Brazilian cities has improved since gasohol started replacing petrol and diesel as a fuel.

5) And now the bad news: a lot of land but not many labourers are needed to grow sugar cane for fuel — so food has got more expensive and many people have lost their jobs in farming.

Alternative fuels — yeah, no bad petrol vibes here, man...

Before you encourage everyone to rush out and pour vodka and slimline tonic into their fuel tanks, remember that traditional petrol and diesel engines often need to be converted before they can run on alternative fuels. One exception is biodiesel — you can usually put that straight in a normal diesel engine.

Revision Summary for Section 2.9

Why does it cost so much to go to London on the train? That's the transport question which puzzles me most, but they probably won't ask that in your exam. They might ask things a bit like this though:

1)* What's the speed of a bus which travels 6 metres in 2 seconds?

2)* A tiger is running at 18 m/s. How far will the tiger get in:

      a) 2 s,   b) a minute?

3)* Rosie paddles her canoe across a lake at a velocity of 2 m/s eastwards. The lake is 500 m wide. How long will it take Rosie to cross the lake?

4) Rosie's friend George is paddling his canoe northwards up the lake, also at 2 m/s. Does George have the same velocity as Rosie?

5)* Calculate the acceleration of a goat which speeds up from 0 to 12 m/s in 5 seconds.

6)* Ella is cycling to work at a speed of 18 m/s. Calculate her deceleration if it takes her 6 s to slow down to a stop.

7)* After winning the 100 m sprint on sports day, Brian slowed down from 7 m/s to 1.5 m/s over 2 seconds. Calculate his deceleration.

8) What two factors influence a driver's thinking distance?

9) Give four factors that might affect a driver's reaction time.

10)* Professor Higgins is driving at 15 m/s. He sees a child run into the road 70 m in front of his car. Professor Higgins' reaction time is 0.5 s. Calculate his thinking distance.

11) List six factors which would increase a car's braking distance.

12) Why are vehicle collisions more likely in fog? What could drivers do to reduce the risk?

13) It is illegal to drive a car with less than 1.6 mm tread on the tyres. What's the reason for this law?

14) Describe what an MOT test is for.

15) How does drinking alcohol affect drivers?

16) Name three ways that technology has improved the safety of drivers and passengers in vehicles.

17) Give five measures that are intended to improve road safety for pedestrians. Explain why each measure should help keep pedestrians safer.

18) Why is crude oil useful to the transport industry?

19) What is a hydrocarbon?

20) Why do hydrocarbons make good fuels?

21) Name the products formed when a hydrocarbon burns in plenty of air.

22) Write a balanced symbol equation for the complete combustion of C_7H_{16}.

23) Describe the energy transfers that take place in a car engine.

24) What happens to the wasted heat from a car engine?

25) Why might the combustion of a hydrocarbon be incomplete?

26) Name the possible products of the incomplete combustion of a hydrocarbon.

27) Explain why incomplete combustion in a vehicle engine makes the engine less efficient.

28 Describe two ways in which incomplete combustion of hydrocarbons is a health hazard.

29) Name three fossil fuels commonly used as fuels for transport.

30) What are the units of energy?

31) What does the 'energy content' of a fuel mean?

32) Explain why scientists are developing alternatives to petrol and diesel to use in cars, buses, etc.

33) Apart from energy content, give four other factors which are important in a new fuel.

*Answers on p.148.

Electromagnetic Waves

Electromagnetic radiation is things like light and X-rays. It's made up of waves and is a bit odd, frankly.

Waves Transfer Energy but NOT Matter

Electromagnetic waves (EM waves) transfer energy from one place to another
without moving any matter (stuff). They're all over the place:

A radio transmitter is a source of radio
waves. These waves carry energy from
the transmitter to people's radio aerials.

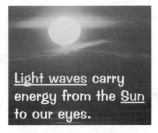

Light waves carry
energy from the Sun
to our eyes.

You can't see EM waves like you can see waves at the seaside. But they're similar in many ways.
All waves have a few basic features that you need to know about:

1) WAVELENGTH is the distance from one peak to the next.
   It can be measured in metres, mm or any other unit of length.

2) FREQUENCY is the number of waves that the source produces
   per second. (A source is something like a radio transmitter.)

3) Frequency is measured in hertz (Hz).
   1 Hz = 1 wave per second.

4) The frequency is all that makes one type of EM wave different
   from another. E.g. light waves are basically the same as radio waves — but with a different frequency.

wavelength

The Higher the Frequency, the More Energy the Waves Have

EM waves with different frequencies have different properties — and different uses.
There are seven basic types — shown below with increasing frequency from left to right:

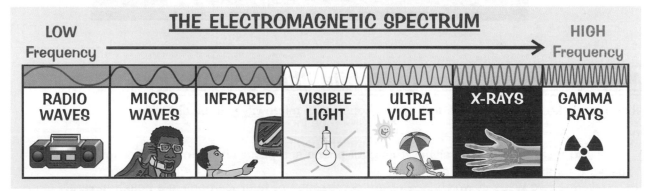

THE ELECTROMAGNETIC SPECTRUM

LOW Frequency ———————————————→ HIGH Frequency

| RADIO WAVES | MICRO WAVES | INFRARED | VISIBLE LIGHT | ULTRA VIOLET | X-RAYS | GAMMA RAYS |

1) The higher the frequency of the waves, the more energy they have — so, for example, gamma rays
   have much more energy than radio waves.

2) In general, the more energy the wave has, the more dangerous it can be — see p.106.

EM Waves — not so much fun for paddling in...

The wavelength and frequency of EM waves are related — the higher the frequency of a wave, the
shorter the wavelength. (You can see this in the diagram of the electromagnetic spectrum above.)

Radio Waves and Microwaves

Waves with fairly <u>low frequencies</u> are often used for <u>communication</u> — radio, TV, mobile phones etc.

Low-Frequency Waves are Used for Communication

1) When EM waves hit a substance, three things can happen. The waves might:
   - be <u>transmitted</u> — just <u>pass through</u> the substance, like light through glass
   - be <u>reflected</u> — <u>bounce back</u>, like light off a mirror
   - be <u>absorbed</u> — the wave's <u>energy</u> is transferred to the substance, like microwaves heating food.
2) EM waves above a certain frequency are <u>absorbed</u> quite a bit by the <u>Earth's atmosphere</u>. That's why we use waves at the <u>low-frequency</u> end of the spectrum for broadcasting signals over long distances.

Radio Waves are Used for Radio and TV Signals

There's a range of wavelengths within the 'radio' part of the spectrum. Radio waves can be transmitted round the world in <u>different ways</u>, depending on their <u>wavelengths</u>. (Remember, shorter wavelengths mean higher frequencies.)

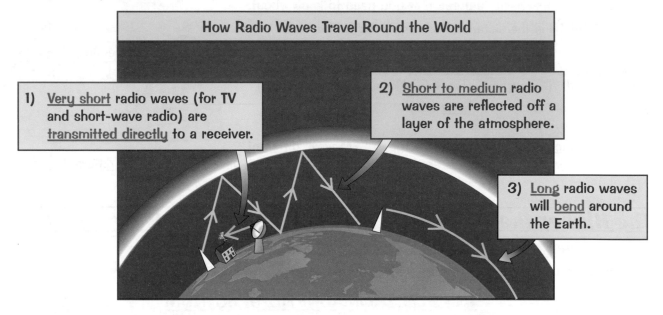

How Radio Waves Travel Round the World

1) <u>Very short</u> radio waves (for TV and short-wave radio) are <u>transmitted directly</u> to a receiver.

2) <u>Short to medium</u> radio waves are reflected off a layer of the atmosphere.

3) <u>Long</u> radio waves will <u>bend</u> around the Earth.

Microwaves are Used in Mobile Phone Networks

When you make a mobile phone call, your phone emits <u>microwaves</u> in all directions.
And then...

1) A nearby <u>mast</u> receives the signal from your phone and passes it on to a central system called the <u>mobile telephone exchange</u>.
2) The exchange passes on the signal to the phone mast <u>nearest to</u> the phone you're calling...
3) ...where it's <u>transmitted</u> via microwaves to the person you're talking to.
4) The are masts (or <u>base stations</u>) all over the country, so you can make a call wherever you are (unless you happen to live in a small village in Cumbria).

Microwaves — when the Queen can't really be bothered...

Remember, electromagnetic waves come in many <u>different frequencies</u> (and wavelengths) and they behave differently in different substances. That's why EM waves have such a wide range of <u>uses</u> — radio and microwaves for communication, light for seeing things, X-rays for, well, X-rays. Good eh.

Infrared and Light

Light's great for <u>seeing things</u>, but it has other uses.

Infrared is Used for Very Short-Range Communication

<u>Infrared waves</u> are used in <u>remote controls</u>, e.g. for TVs and DVD players.

1) An infrared <u>signal</u> transmits instructions from the control to the appliance.

2) Infrared waves are quickly <u>absorbed</u> by most materials. That's why TV remote controls only work if there's nothing blocking the path between them and the TV.

Fibreoptic Cables Carry Light Waves

<u>Broadband</u> internet uses <u>fibreoptic cables</u> to carry <u>data</u> (vast amounts of it) as pulses of <u>light</u>.

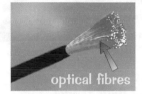

optical fibres

1) One fibreoptic cable contains a <u>bundle</u> of many individual <u>optical fibres</u>.

2) <u>Optical fibres</u> can carry signals over long distances using <u>light waves</u> (see below).

Optical Fibres Work by Repeated Reflections

The diagram shows a single optical fibre. This is how it carries signals:

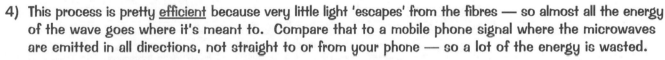

repeated reflections

inner core

plastic sheath

outer layer

1) A light wave enters one end of the fibre.

2) It bounces off the wall of the <u>inner core</u> over and over again...

3) ...until it emerges at the other end.

4) This process is pretty <u>efficient</u> because very little light 'escapes' from the fibres — so almost all the energy of the wave goes where it's meant to. Compare that to a mobile phone signal where the microwaves are emitted in all directions, not straight to or from your phone — so a lot of the energy is wasted.

5) The reason it works so well is that light waves hit the sides of the inner core at a shallow enough <u>angle</u>:

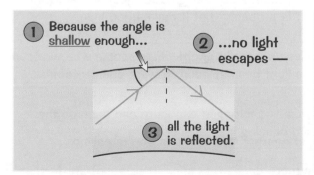

1) Because the angle is <u>shallow</u> enough...

2) ...no light escapes —

3) all the light is reflected.

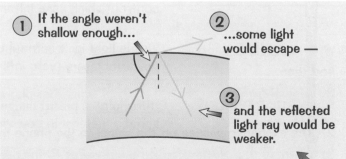

1) If the angle weren't shallow enough...

2) ...some light would escape —

3) and the reflected light ray would be weaker.

6) It's important that optical fibres aren't <u>bent</u> too much — the light might end up at the wrong angle.

Fibreoptics — cables with a high bran content...

Optical fibres have many important uses — <u>telephone networks</u> and <u>cable TV</u> for starters (the 'cable' in cable TV is a fibreoptic cable). They're also used in <u>endoscopes</u> — the tuby things doctors sometimes use to see inside a patient's body without cutting big holes. Oh, and in fake plastic Christmas trees.

Risks and Benefits of Using EM Waves

Radios and telephones are jolly useful and pretty safe, but some EM radiation can be harmful.

Communication Devices Have Changed Society

1) A hundred or so years ago, the only way to communicate with people on the other side of the world was by letters carried on ships and trains — and it took weeks.

2) Now, with fibreoptic cables, satellites and whatnot, we can communicate with people almost anywhere in seconds — by landline phone, mobile phones or email. This has brought many other changes —

We can be better informed about events around the world...

...and the internet makes finding and sharing information easier.

People can work from home (or while they're on the train, even). Homeworkers cause less pollution than people who drive their cars to work every day...

...but they can become socially isolated if they work alone all the time.

We can also trade easily with people all over the world. That's great for people who can sell more products or services, but there are downsides, e.g. international trading can make fraud harder to detect.

Some EM Waves Can be Harmful to People

Generally, the higher frequency (so more energy) an EM wave has, the more harm it can do in your body.

INCREASING FREQUENCY ↓

RADIO WAVES	As far as we know, radio waves just pass through the body harmlessly.
MICROWAVES	Some wavelengths of microwaves are absorbed by water. If that happens in your body, it heats your living cells up — which could be bad for you. Mobile phone networks use microwaves (see p.104), and some people think that using your mobile a lot, or living near a mast, might increase the risk of tumours. There isn't any conclusive proof either way yet.
INFRARED	A large dose of infrared could cause burns or heatstroke — but these risks are easily avoidable. The most dangerous thing about your TV remote control (which uses infrared) is probably that it makes you lazy.
VISIBLE LIGHT	Visible light isn't harmful unless it's really bright. E.g. people who work with powerful lasers (very intense light beams) need to wear eye protection.
ULTRAVIOLET	Ultraviolet is pretty dangerous — prolonged exposure can damage the genetic material in your cells, causing skin cancer.
GAMMA / X-RAYS	X-rays and gamma rays are more dangerous still. They penetrate further into the body than UV and can both cause cancers.

For long-distance calls, always shout loudly into the phone...

There's no point being paranoid about the risks of 'modern technology'. It's better to use your mobile phone to call 999 in an emergency than to let your house burn down while you run two miles to the nearest phone box, because you're worried about the possible health risks of microwaves on your brain.

Waves and Astronomy

A lot of electromagnetic waves reach Earth from space. Astronomers can detect them with <u>telescopes</u> and use the data they collect to find out about the Universe (and life, and everything).

When a Wave Source Moves the Frequency Seems to Change

If you observe a source of waves, something funny happens if that wave source <u>moves</u>. You can spot the effect in the <u>sound waves</u> from a car which <u>zooms past</u> at high speed while you're standing still:

1) As the car moves <u>towards you</u>, listen for the <u>pitch</u> of the sound its engine makes.

2) As the car moves <u>away from you</u>, the engine noise will <u>sound lower pitched</u> — the sound waves seem to have a <u>lower frequency</u>. It sounds a bit like vrerrrrr-ooom. This is why:

3) So if you stood by a busy road with your eyes shut, you could tell when cars were <u>moving away</u> from you, just from the <u>drop in frequency</u>.

4) You could also tell which cars were moving faster — the <u>faster</u> they're moving, the <u>greater the drop</u> in frequency when they move away.

Light from Other Galaxies is Red-Shifted

1) Most of the weird and wonderful objects in space emit EM waves at a whole variety of frequencies — and various different telescopes are used to detect them.

2) From all their observations, astronomers have found that the light waves coming <u>from distant galaxies</u> are all at <u>slightly lower frequencies</u> than expected — they're <u>shifted</u> towards the <u>red end</u> (the low-frequency end) of the visible spectrum (see p.103). This effect is called the <u>red-shift</u>.

3) <u>Measurements</u> of the red-shift suggest that <u>all the galaxies</u> are <u>moving away from us</u> very quickly.

The Universe Seems to be Expanding

1) <u>More distant</u> galaxies have <u>greater</u> red-shifts than nearer ones.

2) This means that more distant galaxies are <u>moving away</u> from us <u>faster</u> than nearer ones.

3) This provides evidence that the whole Universe is <u>expanding</u>.

Red-shift, I asked you nicely before — now just move...

You only notice a drop in frequency from things <u>moving relative to you</u> and pretty <u>fast</u> — like cars (or galaxies). And remember, the frequency only <u>seems</u> to change. For people <u>inside</u> the car, the noise of its engine <u>doesn't</u> change as they go past you — because the frequency <u>hasn't actually</u> changed. Odd.

Revision Summary for Section 2.10

Electromagnetic waves — I'll admit they're not a barrel of laughs, but they are useful. For a start, without the light waves being reflected off this page, you wouldn't be able to read all these questions. That would be a mighty shame, because you wouldn't be able to try answering them all. And unless you can answer them all, you probably need to do a bit more revision — lucky you.

1) What do electromagnetic waves move? What don't they move?
2) Draw a diagram showing what wavelength means.
3) Define frequency.
4) What are the units of frequency?
5) How are ultraviolet waves different from visible light waves?
6) Write out the electromagnetic spectrum, putting the lowest frequency waves on the left.
7) How are the energy and frequency of EM waves related?
8) What three things might happen when a light wave reaches the surface of some water?
9) Explain how radio waves allow us to communication over long distances.
10) What kind of EM waves are emitted by mobile phones?
11) Describe how satellites are used in mobile phone networks.
12) What kind of EM waves are emitted by TV remote controls?
13) Explain why a TV remote control won't work if you're hiding behind the sofa when you try to use it.
14) What is: a) a fibreoptic cable, b) an optical fibre?
15) Sketch and label a diagram of an optical fibre.
16) What kind of EM waves are used in optical fibres?
17) Why won't optical fibres work properly if they're bent too much?
18) Give three uses of fibreoptic cables.
19) Name three methods of long-distance communication that rely on electromagnetic waves.
20) Outline how people are affected in their jobs by:
    a) the internet,
    b) mobile phones.
21) Describe one way in which modern communication devices can be beneficial for the environment.
22) What kinds of EM waves are generally the most harmful in the human body? Explain why.
23) How do radio waves affect the human body?
24) What happens to the cells in your body if they absorb microwave radiation?
25) How dangerous are the EM waves emitted by remote controls for DVD players?
26) Why are lasers potentially dangerous?
27) Sunbeds emit ultraviolet radiation. Explain why using sunbeds for long periods of time is unwise.
28) Which types of EM radiation are known to cause cancer?
29) Sarah is chatting to her friend outside the corner shop when an ambulance drives by, siren blaring. What change might Sarah notice in the sound of the siren as the ambulance passes?
30) What might she notice if the ambulance drove back in the opposite direction, even faster?
31) Why are telescopes so great?
32) How can astronomers tell that a certain galaxy is moving away from Earth?
33) Explain how astronomers can work out which galaxies are moving away fastest.
34) Why do scientists think that the Universe is expanding?

Following Standard Procedures

This section is all about how scientists work when they carry out <u>practical tasks</u>, how they <u>record results</u>, how they <u>interpret results</u>, basically how they <u>carry out their jobs</u>. But first things first, they have to be able to follow a <u>procedure</u>. Whether making acids or measuring giraffes, scientists follow '<u>standard procedures</u>' — clear instructions describing exactly how to carry out these practical tasks.

Standard Procedures Mean Everyone Does Things the Same Way

Standard procedures are <u>agreed methods of working</u> — they are chosen because they're the <u>safest</u>, <u>most effective</u> and <u>accurate methods</u> to use. Standard procedures can be agreed within a company, nationally, or internationally.

Standard procedures are used for a wide range of things, e.g.
1) <u>Taking measurements</u> (e.g. measuring the density of boats).
2) <u>Preparing and purifying compounds</u> (e.g. making medicines).
3) <u>Monitoring changes</u> (e.g. monitoring water quality).

There are Seven Steps to Following a Standard Procedure

When following a standard procedure, you need to:
1) <u>Read the procedure</u> and check you <u>understand</u> everything.

2) Complete a <u>health and safety</u> check of your <u>working area</u>.

3) Complete a <u>risk assessment</u> for the activity.

There aren't that many risks associated with this experiment (there's no <u>dangerous chemicals</u> or <u>risk of explosion</u>). The biggest risk is the possibility of dropping <u>heavy weights</u> onto your feet.

4) <u>Collect the equipment and materials</u> you need, and set out your working area.

5) Follow the instructions <u>one step at a time</u>.

6) Select instruments that give <u>appropriate precision</u>, and use them to make <u>accurate observations</u> or <u>measurements</u>.

7) Identify possible sources of <u>error</u> and <u>repeat observations</u> and <u>measurements</u> where necessary to improve <u>reliability</u>.

Densicorp Standard Procedure for Measuring the Density of Materials:
1) Ensure test specimens have a regular shape (e.g. cubes or cylinders).
2) Measure the specimen's mass to the nearest 0.01 g.
3) Measure the specimen's dimensions to the nearest mm.
4) Calculate the volume of the specimen.
5) Calculate the density of the specimen using the formula: density = mass ÷ volume.

You'll need a <u>top pan balance</u>, a <u>ruler</u> and some <u>materials</u> to measure.

Your balance will need to be able to <u>accurately</u> measure to the <u>nearest 0.01 g</u>, and your ruler to the <u>nearest mm</u>.

Before packing away all your equipment have a look over your <u>results</u> — if any look <u>out of place</u> repeat the experiment and have a <u>think</u> about why they might have been <u>wrong</u>.

Tamoto techkup — a sauce of error...

There are so many reasons why you should follow <u>standard procedures</u>. They're <u>tried and tested</u>, so you're more likely to get <u>good results</u> and less likely to <u>injure yourself</u>. They should have them for everything, like getting dressed — there's more 'trouser-related accidents' every year than you'd think.

Handling Scientific Equipment

When carrying out any experiment, one of the most important things is using the right equipment for the job at hand. You wouldn't get very far if you tried boiling water in a china tea cup using a cigarette lighter, and testing the temperature of the water with your fingers — trust me, I've been there.

Standard Lab Equipment — Know Your Stuff

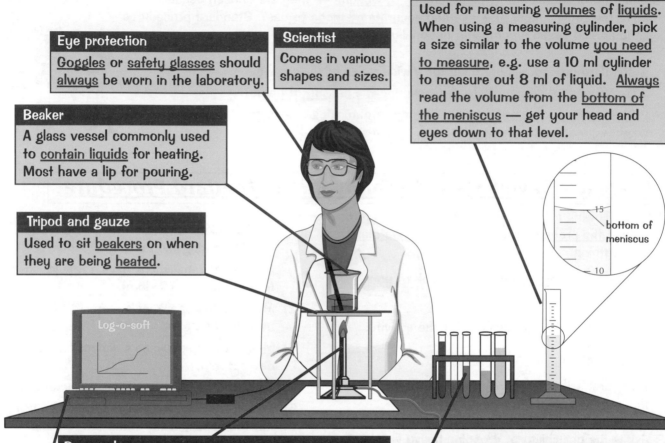

Measuring cylinder
Used for measuring volumes of liquids. When using a measuring cylinder, pick a size similar to the volume you need to measure, e.g. use a 10 ml cylinder to measure out 8 ml of liquid. Always read the volume from the bottom of the meniscus — get your head and eyes down to that level.

Eye protection
Goggles or safety glasses should always be worn in the laboratory.

Scientist
Comes in various shapes and sizes.

Beaker
A glass vessel commonly used to contain liquids for heating. Most have a lip for pouring.

Tripod and gauze
Used to sit beakers on when they are being heated.

15
bottom of meniscus
10

Bunsen burner
Used to heat substances in the laboratory. They have an air hole at the bottom which can be opened and closed. When the valve is fully closed the flame burns orange (this is known as the safety flame), but when closed the flame is blue. The Bunsen should always be lit with the air hole open. And remember — always use a heatproof mat.

Test tube rack, test tubes and boiling tubes
Test tubes are long thin glass tubes with a U-shaped bottom. They are used to hold small amounts for heating, though boiling tubes (larger versions of test tubes) are usually favoured for heating liquids.

Data logger
Data loggers accurately record changes in things like temperature, pH and salinity over time. You need to be able to set up a data logger (which may involve attaching different probes), transfer the data to a computer and be able to calibrate it. E.g. to calibrate a temperature probe, dip it into freezing water and then into boiling water. To calibrate pH or salinity, data loggers are usually supplied with a set of standard solutions with known values.

Beaker, Bunsen — we're just missing Kermit and Fozzie Bear...

It's all very well knowing how to use this gear, but you've also got to be able to use it safely. Remember the basics, like not eating or running and generally not acting like an idiot in the lab — if your memory needs refreshing have a look back over Section 1.2 (pages 6-11).

Handling Scientific Equipment

In microbiology, everything is pretty small (the clue's in the name). A lot of the equipment you'll need to use for microbiology makes handling and seeing small things a lot easier.

Equipment for Microbiology

Microscope
Used to view microorganisms and other things which are too small to view with the naked eye.

Mounted needle
Used to prepare glass slides to avoid getting grubby fingerprints all over them.

Glass slides and cover slips
Before samples can be viewed they need to be placed onto a glass slide. A cover slip is then placed on top to hold the sample in place.

Sample bottles
Bottles to put samples in.

Pipette
Used to add small volumes of a liquid, drop by drop.

Petri dish
Shallow plastic or glass dish used to grow microorganisms in. They're filled with a gel (called agar), which contains the nutrients needed for microorganisms to grow.

Stain
To make some samples easier to view they are stained. A common stain used in the lab is methylene blue.

Inoculating loop
Used to transfer samples, e.g. bacteria, onto agar plates.

Assay discs
Small circles of filter paper which are soaked in different chemicals and then placed on top the agar. This is done to investigate the effect of different chemicals on microorganisms.

Incubator
When microorganisms are cultured (grown) they must be kept at a certain temperature. Incubators are used to keep them at this temperature. (Water baths can sometimes also be used).

Autoclave
Once the experiment has finished the microorganisms must be disposed of safely. An autoclave is like a large, very hot pressure cooker that kills the microorganisms using high temperatures.

Assay old chap, what a ruddy good page. Brandy anyone...

One of the most important things when handling microorganisms is to use aseptic techniques — these stop you spreading nasty microorganisms everywhere. There's more on aseptic techniques on p.115.

Handling Scientific Equipment

Microscopes and Petri dishes wouldn't be much use for flame tests, so there's even more specialised equipment needed for chemical analysis.

Equipment for Chemical Analysis

Burette

Used to dispense a measured volume of liquid. Their most common use is in titrations (see p.120). As with measuring cylinders (see p.110) you need to read the volume from the bottom of the meniscus. You should fill pipettes and burettes to about 3 cm above the desired amount, then carefully drop the level down to what you need. It's also important to clamp the burette properly so that it doesn't fall over and smash.

Volumetric flask

Used to accurately measure a large volume of liquid. They are particularly useful when preparing solutions of a specified concentration.

Conical flask

These are used when the contents need to be swirled during an experiment, e.g. when reaching the end point of a titration.

Pipette and filler

Used to deliver a measured volume of liquid.

Sample tubes

Used to hold chemical samples.

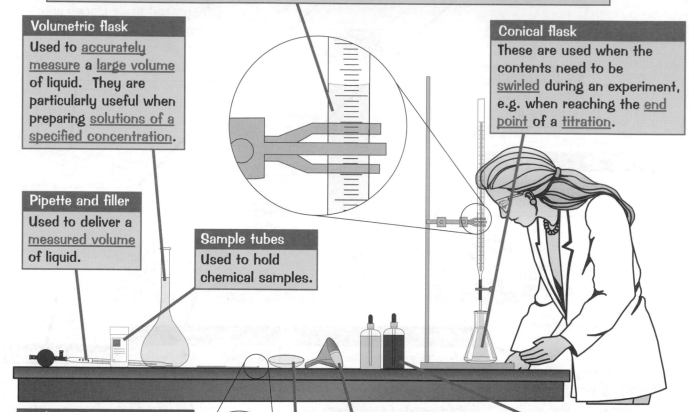

Nichrome wire

Piece of wire with a handle at one end and a small loop at the other. Usually used in flame tests.

Watch glass

A circular piece of glassware shaped like a small bowl. It's used in experiments to view an evaporating liquid.

Filter funnel

Used for separating solids from liquids (with filter paper folded up and placed inside the funnel). Also useful for filling things with a narrow neck (e.g. burettes and volumetric flasks).

Indicators

Indicators provide a clear colour change when a reaction takes place. E.g. they're used for determining the end point of a titration.

Top pan balance

Used to weigh out amounts of a substance. They can measure with an extremely high degree of precision. It's important to 'zero' the balance before use (press the T (tare) button when there's nothing on it). Balances need to be calibrated (they usually come with a set of weights of known mass).

Pipettes — aren't they a girl band from Brighton...

The same old safety rules apply, but be careful — analytical chemistry uses loads of hazardous chemicals.

Handling Scientific Equipment

Like all good things in life, this small section on scientific equipment must come to an end. But wipe away those tears — this page on equipment for materials testing is going to be a right hoot. And as if that wasn't enough, the next few pages are all about investigating living organisms — yay.

Equipment for Materials Testing

Clamp and stand
Large metal stand with a clamp. The clamp can be moved up and down the stand. In materials testing clamps and stands are mostly used for hanging things off to test their strength or elasticity.

Material to be tested
This could be anything. You may need to determine the conductivity of a piece of metal, or the strength of a piece of wire.

Mass hanger and masses
Used for testing the strength of materials. There are different sized masses depending on what you're testing. It's important to pick the right mass for the material you're testing, e.g. it'd be daft to test the strength of cotton thread with a 1 kg mass.

Displacement can
Displacement cans are filled up to just below the arm. The material to be tested is then placed into the can. The amount of water it displaces can be collected and measured.

Power pack
Allows the user to change the voltage of the mains supply to suit the needs of the experiment.

Voltmeter
Used to measure the voltage across a component. Remember, voltmeters must always be connected in parallel.

Ammeter
Measures the current flowing through a circuit. Ammeters should always be connected in series.

Wires and electrodes
These are used to connect everything together and make the test circuit complete.

Components
Materials scientists also use a range of components including fixed capacity and variable resistors, filament lamps, light-emitting diodes and thermistors.

Clamp... tweezers... stat...

Well, after all of that you should be pretty familiar with the standard equipment used by scientists every day. The more you use equipment like this, the more accurate your experiments will become. You'll also become better at using less familiar equipment. It's a bit like riding a bike — the more you ride a bike, the better you get and the better you'll be at riding something less familiar, like a horse. OK, so it's not really like riding a bike, but people should ride bikes more often.

Investigating Living Organisms

Microbiologists use <u>microscopes</u> to study things that are <u>too small</u> to see with the <u>naked eye</u>. This can include really, really small <u>organisms</u> (like bacteria) or the <u>cells</u> and <u>tissues</u> that make up <u>larger organisms</u>.

Ten Easy Steps to Setting Up a Light Microscope

1) <u>Always</u> carry your microscope by the <u>handle</u>.
2) If it has a <u>built-in light</u>, plug it in and switch it on.
3) If your microscope has a <u>mirror</u>, place it near a <u>lamp</u> or a <u>window</u>, and angle the mirror so light shines up through the <u>hole in the stage</u>.

> Don't reflect <u>direct sunlight</u> into the microscope — it could <u>damage your eyes</u>.

4) Place a <u>slide</u> on the <u>stage</u>, and clip it in position.
5) Select the <u>lowest</u> powered <u>objective lens</u>.
6) <u>Before</u> looking down the microscope, <u>position</u> the objective lens <u>just above</u> the slide (do this by turning the <u>focusing knob</u>, which raises and lowers either the objective lens or the stage).
7) <u>Look</u> down the <u>eyepiece</u>, and carefully start to <u>focus</u> by turning the focusing knob.
8) Focus until you get a <u>clear image</u>.
9) If you need to look at the slide with <u>greater magnification</u>, switch to a <u>higher powered</u> objective lens (a longer one). With some microscopes you can also swap the <u>eyepiece lens</u>.
10) <u>Refocus</u> the microscope (repeat steps 7 and 8).

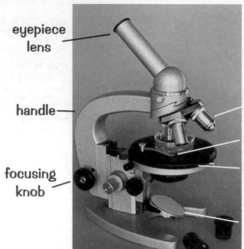

Labels: eyepiece lens, handle, focusing knob, objective lens, clip, stage, mirror or built-in light

> Always turn the <u>focusing knob</u> so that the objective lens is <u>moving away</u> from the slide — this is so the lens and slide don't <u>crash together</u>. Some microscopes have a <u>coarse</u> focusing knob and a <u>fine</u> focusing knob. Use the coarse knob to make <u>large adjustments</u>, and the fine knob to make <u>smaller ones</u>.

Samples Need to be Prepared Before Investigation

You can't just slap a piece of tissue underneath a microscope — it has to be on a slide.

1) Use a pipette to put <u>one drop</u> of <u>mountant</u> (a clear, gloopy liquid) in the middle of the slide — this <u>secures</u> the sample in place. Sometimes <u>water</u> can be used.
2) Use <u>forceps</u> to place your sample on the <u>slide</u> (e.g. a <u>hair</u> for <u>forensic examination</u>).
3) Make sure the mountant is <u>holding</u> the sample in place, and it's positioned so it will <u>all</u> be <u>under the cover slip</u>.
4) Sometimes a drop of <u>stain</u> (e.g. <u>methylene blue</u>) is added to make the samples <u>easier</u> to see under a microscope.
5) Place the cover slip at <u>one end</u> of the sample, holding it at an angle with a <u>mounted needle</u>.
6) Carefully <u>lower</u> the cover slip onto the slide. Press it down <u>gently</u> with the needle so that no <u>air bubbles</u> are <u>trapped</u> under it.

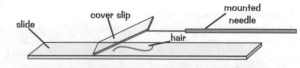

Labels: slide, cover slip, hair, mounted needle

> Always handle slides and cover slips by their edges to avoid <u>finger marks</u>.

Cover, slip — I thought this was science, not cricket...

Nobody knows exactly <u>who</u> invented microscopes or <u>when</u> they were invented, but it's thought to have been some time in the <u>16th century</u> — lots of people claim to have invented them (including my uncle Charlie, but he's slightly crazy). These days we've got some pretty cool microscopes, e.g. <u>electron microscopes</u> — these produce <u>3D images</u> and have <u>magnifications</u> beyond your wildest dreams.

Section 3 — Developing Scientific Skills

Investigating Living Organisms

In their investigations, microbiologists aim to get the best results possible and to stay out of harm's way whilst they're doing it — they achieve this using aseptic techniques.

Microorganisms are Everywhere

Microorganisms are everywhere — in the air, on our hands, on laboratory benches and in hospitals. There are useful microorganisms, like the ones that make bread, beer, yoghurt and antibiotics (see pages 47-48). There are also harmful microorganisms — ones that can make us ill.

Microbiologists often isolate and grow colonies of microorganisms for investigation. Culturing (growing) microorganisms can be dangerous, so scientists need to:

1) Make sure they don't contaminate the laboratory with the microorganisms they culture.

2) Make sure their work doesn't become contaminated with microorganisms from the environment.

Aseptic Techniques Prevent Contamination

Aseptic techniques are standard procedures used by microbiologists to prevent contamination. The techniques involve creating a clean, contained environment to culture microorganisms in.

1) Sterilise all equipment before and after use.

2) Keep samples containing microorganisms in sample bottles with lids.

3) When opening a sample bottle to use it, close it again as soon as possible.

4) Pass the tops of sample bottles through a Bunsen flame whenever lids are removed.

5) Don't put lids down on benches — hold them with your little finger or your other hand.

6) Don't open Petri dishes until you are ready to use them.

7) Don't put any equipment that comes into contact with microorganisms down on benches.

8) Seal agar plates with sticky tape.

9) Before the plates are incubated, label them with your name, the date and what you've put on the plate.

10) Don't open agar plates once they have been sealed.

11) Dispose of cultures safely — usually done by pressure sterilising in an autoclave.

In large commercial laboratories they employ chickens to sit on plates and incubate them.

There are even Microorganisms in Milk

To show that microorganisms really are everywhere and to put your aseptic techniques to the test, why not try to culture the microorganisms in milk.

1) Heat a wire loop in a Bunsen flame until it glows red hot (to sterilise it), and then let it cool for 30 seconds. Don't allow the loop to touch anything.

2) Dip the cooled loop into a sample of milk.

3) Remove the lid of the Petri dish and streak the wire loop across the surface.

4) Replace the lid on the Petri dish and secure it with strips of sticky tape.

5) Label the dish with your name, the date, and what's on the agar.

6) Incubate the plate for 48 hours at 25 °C.

7) Examine the plate without opening it, and dispose of it safely.

Working with sewage requires a septic technique...

It's a bit dull and some of it's just common sense, but it's still important. Scientists in industry need to use aseptic techniques all the time — and they're crucial in hospitals.

Investigating Living Organisms

Some microorganisms are just plain <u>nasty</u> — luckily we have <u>antimicrobial agents</u> to kill them. Other microorganisms make <u>tasty things</u> (plain or flavoured). Don't take my word for it though — you can try them both using... yep, you guessed it — some more <u>standard procedures</u>.

Antimicrobial Agents Inhibit the Growth of Microorganisms

<u>Antimicrobial agents</u> (like <u>antiseptics</u>, <u>disinfectants</u> and <u>antibiotics</u>) <u>kill</u> or <u>prevent</u> the <u>growth</u> of microorganisms. You can compare the <u>effectiveness</u> of different antimicrobial agents:

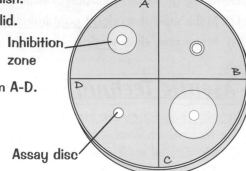

1) Use a pipette to put five drops of <u>bacterial culture</u> into a Petri dish.

2) Pour 20 cm³ of molten agar into the Petri dish and replace the lid.

3) <u>Mix</u> the contents by sliding the dish <u>gently</u> over the bench.

4) Leave the plate to <u>set</u>.

5) Mark out <u>four quadrants</u> on the base of the dish, and <u>label</u> them A-D.

6) Using <u>forceps</u>, dip a 5 mm assay disc into <u>distilled water</u> and place the disc on the surface of the agar, in the middle of quadrant D — this is the <u>control</u>.

7) Dip three assay discs into three <u>different antiseptics</u> and add to the other quadrants — <u>remember</u> to note <u>which antiseptic</u> went in <u>which quadrant</u>.

8) Tape down the lid, and <u>incubate</u> the dish at 25 °C for 48 hours.

9) Measure the <u>zone of inhibition</u> around each disc — you could stand the dish on graph paper and count the squares beneath each zone, or just measure the diameter and calculate the area of the circle).

> The <u>zone of inhibition</u> is the <u>clear area</u> around an assay disc where <u>no microbes have grown</u>. The <u>larger</u> the inhibition zone, the <u>more effective</u> the antimicrobial agent.

Microorganisms Can be Used to Make Yoghurt

Some microorganisms help produce <u>useful products</u>, e.g. beer, wine, bread and a whole host of dairy products are all made using microorganisms. Ever fancied making your own yoghurt — then read on.

1) Pour 250 cm³ of **UHT** milk into a <u>sterile beaker</u>.

2) Place the beaker in a <u>water bath</u> at 37 °C and <u>stir gently</u> until the milk reaches the same temperature.

> The <u>microorganisms</u> that make yoghurt will <u>die</u> if they get <u>too hot</u>, and they won't grow very fast if they're too <u>cold</u>.

3) Add some '<u>yoghurt starter culture</u>' and stir — you can buy special starter cultures, or just use a tablespoon of <u>ordinary live yoghurt</u>.

4) <u>Cover the beaker</u> with foil and leave it in the water bath for <u>24 hours</u>.

5) Place the beaker in a bowl of <u>cold water</u> and stir until <u>smooth</u>.

6) Put the beaker in the <u>fridge</u> for a few hours for the yogurt to <u>thicken</u>.

Aunty Microbial — she's married to uncle Bob...

<u>Zones of inhibition</u>, <u>agents</u> killing stuff — this sounds more like a page from a book written by a former member of the SAS than how to develop your <u>scientific skills</u>. I'd love to tell you more secrets about yoghurt making, but I'd have to kill you. On that note, I think it's time for some <u>chemistry</u>.

Analytical Chemistry

Chemical analysis is important in environmental science. Say lots of fish in a river die suddenly and it's suspected that pollution from a nearby factory is to blame — an analytical chemist could find out exactly what's in the water. Forensic scientists also use chemical analysis, e.g. to detect banned substances in blood or urine samples from athletes or to identify substances present at a crime scene.

Qualitative and Quantitative Analysis

There are two types of chemical analysis:

1) Qualitative analysis tells you what substances are present — this page and the next.
2) Quantitative analysis tells you how much of a substance is present — see p.119-121.

You Can Identify a Compound by the Ions It Contains

One way to identify a compound is to identify its ions (remember, an ion is just an atom that's gained or lost some electrons — see p.70). E.g. if you know a substance has sodium ions and chloride ions in it (and nothing else) you know it's sodium chloride. The two tests that follow are for metal ions.

Add Sodium Hydroxide and Look for a Coloured Precipitate

1) Many metal hydroxides are insoluble — so they precipitate out of solution when formed.
2) Some of these hydroxide precipitates have a characteristic colour.
3) In this test you just add a few drops of sodium hydroxide solution to a solution of your mystery compound, and see what happens.
4) If a precipitate forms, its colour tells you which metal hydroxide you've made — and so what the metal bit of your mystery compound could be...

Metal ion	Colour of precipitate
Calcium, Ca^{2+}	White
Copper(II), Cu^{2+}	Blue
Iron(II), Fe^{2+}	Sludgy green
Lead, Pb^{2+}	White at first. But if you add loads more sodium hydroxide it forms a colourless solution.

Do a Flame Test

Some metal ions produce pretty flames when they burn — the colour of the flame can tell you which metal is present. This is what you do:

1) Prepare a powdered sample of the mystery substance.
2) Get a wire loop and clean it — by dipping it in some hydrochloric acid.
3) Dip the loop into the sample, then hold the end in a blue Bunsen flame.

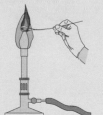

Metal ion	Colour of flame
Sodium, Na^+	Yellowy orange
Potassium, K^+	Lilac
Calcium, Ca^{2+}	Brick red
Copper(II), Cu^{2+}	Blue-green
Lead, Pb^{2+}	Blue

One Test isn't Always Enough...

Watch out — sometimes different metal ions give the same result. For example:

- Aluminium behaves like lead when it reacts with sodium hydroxide.
  So if you only do the sodium hydroxide test, you won't know which it is.

- To be sure, you have to do a flame test as well (lead burns blue, while aluminium doesn't burn with a characteristic colour).

Coloured Precipitate — Purple Rain... (ask your Mum or Dad)

Knowing about all those pretty colours in the flame tests isn't just important for forensic scientists and GCSE Applied Science students — it's also vital for firework-makers. So remember, remember....

Analytical Chemistry

So, imagine you've got a mystery compound and you've already figured out it contains <u>copper</u> ions (using a flame test). You still have to find out whether it's copper <u>carbonate</u>, copper <u>sulfate</u> or what. Luckily, there are more useful tests you can do.

Testing for Carbonates — Use Dilute Acid

<u>Carbonates</u> give off <u>carbon dioxide</u> when added to <u>dilute acids</u>. Here's the method:

1) Put your mystery compound in dilute acid, e.g. <u>dilute hydrochloric acid</u>, and <u>collect</u> any <u>gas</u> given off.

2) Bubble the gas through <u>limewater</u>. If the limewater turns <u>milky</u>, the gas given off is <u>carbon dioxide</u>...

3) ...so your compound contains <u>carbonate</u> ions — CO_3^{2-}.

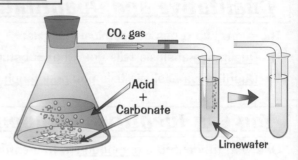

Testing for Sulfates — Hydrochloric Acid then Barium Chloride

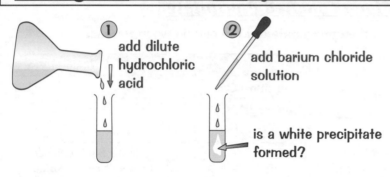

add dilute hydrochloric acid

add barium chloride solution

is a white precipitate formed?

1) Add some <u>dilute hydrochloric acid</u> to a solution of your compound.

2) Then add a few drops of <u>barium chloride solution</u> to the liquid.

3) If you see a <u>white precipitate</u>, there are <u>sulfate</u> ions (SO_4^{2-}) in your compound.

Testing for Chlorides — Nitric Acid then Silver Nitrate

1) Add <u>dilute nitric acid</u> to a solution of your compound.

2) Then add a few drops of <u>silver nitrate solution</u> to the liquid.

3) If you see a <u>white precipitate</u>, there are <u>chloride</u> ions (Cl^-) in your compound.

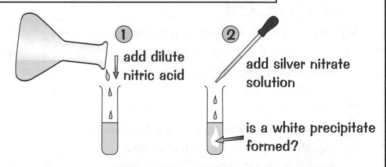

add dilute nitric acid

add silver nitrate solution

is a white precipitate formed?

Drawing Conclusions

The tests above can occasionally be misleading, e.g.:

1) If the <u>carbonate test</u> is positive, it's possible your sample contains hydrogencarbonate ions (HCO_3^-) rather than carbonate ions (CO_3^{2-}).

2) If the <u>sulfate test</u> is positive, your sample might contain hydrogensulfate ions (HSO_4^-) rather than sulfate ions (SO_4^{2-}).

3) The <u>chloride</u> test <u>only</u> works for chloride ions. Hurrah. But... bromide ions (Br^-) produce a <u>cream precipitate</u> that looks quite similar.

Snow White Precipitate — and the Seven Analytical Chemists...

So that's <u>qualitative analysis</u> — finding out what ions you've got and therefore which compound you've got. It's all good clean fun. At least you don't have to <u>memorise</u> what to add to what, and which colours mean which ions. On the next three pages the fun continues with <u>quantitative analysis</u>.

Analytical Chemistry

OK, onto quantitative analysis — concentration is important, for passing exams and for <u>chemistry solutions</u>.

Concentration is Measured in g/dm³ or in mol/dm³

At some point you'll be asked to <u>prepare a solution</u> of a certain <u>concentration</u>.

1) There are <u>two</u> different <u>units</u> of concentration:

 • <u>grams per dm³</u> — g/dm³ ← A '<u>dm³</u>' is a <u>decimetre cubed</u>. It's the same as a <u>litre</u> (1000 cm³).

 • <u>moles per dm³</u> — mol/dm³ ← A <u>mole</u> is just a <u>number</u>. Mol/dm³ is to do with <u>how many molecules</u> of solute there are per litre of solution.

2) With concentration in g/dm³, you're saying what <u>mass</u> of solute is dissolved in <u>every dm³</u> of solvent.

3) E.g. to make 2 litres of 30 g/dm³ solution, you need to dissolve 60 g of solute (2 × 30 g) in water to make up 2 litres of solution. Not too bad. Here's another example, but for a smaller volume of solution:

Example in g/dm³

Dr Brine wants to prepare 500 cm³ of sodium chloride solution at 34 g/dm³ concentration. He has some dry sodium chloride and some water. Describe what he should do.

<u>ANSWER</u>:

He wants to make <u>half</u> a <u>dm³</u> of solution (500 cm³ is half a litre), so he needs 34 ÷ 2 = <u>17 g</u> of solute.

So he should weigh out <u>17 g</u> of sodium chloride and put it in a measuring cylinder, then add <u>water</u> until it he's got <u>500 cm³</u> and <u>stir</u> until it's all dissolved.

You Might Have to Dilute a Solution

You might be given a solution of, say, 3 mol/dm³ concentration and asked to prepare the same solution but in a different concentration, e.g. 0.25 mol/dm³. You'd have to 'water it down'. It's like making orange squash — but with sums and very careful measuring:

Example in mol/dm³

Beth has a <u>2.0 mol/dm³</u> solution of <u>hydrochloric acid</u> and some <u>water</u>. Describe how she could prepare <u>500 cm³</u> of a <u>0.5 mol/dm³</u> solution of hydrochloric acid.

1) Work out how many times the final concentration number goes into the original concentration number. (This tells you how much weaker the solution needs to be.)

2) Divide the FINAL VOLUME by this number — to work out the VOLUME OF ACID you need to use.

3) Subtract the VOLUME OF ACID from the FINAL VOLUME — to get the VOLUME OF WATER needed.

1. 0.5 goes into 2.0 <u>four</u> times.

2. Volume of acid = 500 ÷ <u>4</u> = <u>125 cm³</u>

3. Volume of water = 500 − 125 = <u>375 cm³</u>

<u>ANSWER</u>:

Beth should measure out 125 cm³ of 2.0 mol/dm³ acid in a measuring cylinder and 375 cm³ of water in another measuring cylinder, then add the acid to the water (in a beaker, conical flask or similar).

0.5 moles per decimetre cubed — how many holes can they dig?

With orange squash, you can just keep adding water till it <u>tastes</u> about right. You can't do that in GCSE Applied Science — because the solution you're making might well be poisonous, and because the concentration has to be <u>spot on</u>, not just 'about right'. So <u>measure</u> everything <u>very carefully</u>.

Analytical Chemistry

Imagine that some sulfuric acid has leaked into a river but no one knows how much of it. To find out, you could collect a sample of the river water and find out how concentrated the acid is — by doing a titration.

Titrations are Used to Find Out Concentrations

Doing a titration means finding out what volume of acid is needed to neutralise a known volume and concentration of alkali. (It can be the other way round — finding the volume of alkali needed, if you know about the acid.) Then a quick sum (see next page) is all you need to find the mystery concentration.

Setting Up the Apparatus

1) Use a pipette and pipette filler to measure out the right volume of alkali (often 25 cm³) into a conical flask. Make sure you note down the concentration of the alkali.

2) Add two or three drops of indicator to the alkali.

3) Use a funnel to fill a burette with the acid. Make sure it's filled exactly to the level marked at the top. (Do this with the top of the burette below eye level — you don't want to be looking up if some acid spills over.)

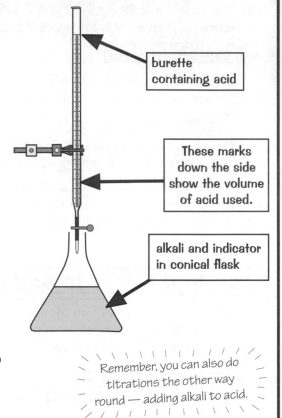

burette containing acid

These marks down the side show the volume of acid used.

alkali and indicator in conical flask

Doing the Titration

1) Using the burette, add the acid to the alkali a bit at a time, giving the conical flask a regular swirl.

2) When you think the end-point (colour change) is about to be reached, go especially slowly — drip by drip.

3) When the indicator changes colour, all the alkali has been neutralised and you should stop adding acid.

4) Read the burette to see what volume of acid you've added to the alkali. Take the reading at eye level, and remember — it's the level of the bottom of the meniscus that matters.

Remember, you can also do titrations the other way round — adding alkali to acid.

Repeat the Experiment

With titrations, as with many of life's most enjoyable experiences, once just isn't enough.

You need to get several consistent readings

To increase the accuracy of your titration and to spot any anomalous results, you need several consistent readings.

- The first titration you do should be a rough titration to get an approximate idea of where the solution changes colour (the end-point).

- You then need to repeat the whole thing a few times, making sure you get (pretty much) the same answer each time (within about 0.2 cm³).

Titrations — not funny, but useful...

Before the end of this unit, you'll be a dab hand at titrations — whether you want to be or not. They're not too tricky really — you just need to make sure that your results are as accurate as possible, which means going slowly near the end-point and then repeating the whole process.

Analytical Chemistry

So you've done the titration — now for the sums.

You Might be Asked to Calculate the Concentration

The whole point of doing a titration is so you can use the results to work out the concentration of an acid or alkali.

Example 1

Brian did the following experiment to work out the concentration of a sample of hydrochloric acid.

He put 20 cm³ of sodium hydroxide solution, concentration 0.1 mol/dm³, in a conical flask. He did a titration and found that 40 cm³ of hydrochloric acid was needed to neutralise the sodium hydroxide.

Work out the concentration of Brian's hydrochloric acid.

The balanced equation for the reaction in Brian's experiment is: $NaOH + HCl \longrightarrow NaCl + H_2O$

1) This means there's one 'HCl' used to neutralise one 'NaOH'.

2) But in Brian's experiment, 40 cm³ of HCl neutralised 20 cm³ of NaOH.
   He needed more acid than alkali — so the acid must have been more dilute.

3 In fact, he needed twice as much acid...

4) ...so its concentration must have been half that of the alkali.
   So the hydrochloric acid had a concentration of 0.1 mol/dm³ ÷ 2 = 0.05 mol/dm³.

Example 2

Professor Plum had a different sample of hydrochloric acid. He found that 30 cm³ of acid was needed to neutralise 25 cm³ of 1 mol/dm³ sodium hydroxide solution. Calculate the concentration of the acid.

The volumes aren't as easy to compare this time. But, you can do it using numbers of moles instead:

Step 1: Work out how many moles of the "known" substance (i.e. the sodium hydroxide) you have, using this formula triangle:

Concentration (in mol/dm³)   Number of moles

$$\frac{n}{c \times V}$$

Volume (in dm³)
One dm³ is a litre

Number of moles = concentration × volume
= 1 mol/dm³ × (25 / 1000) dm³
= 0.025 moles of NaOH

Remember:
1000 cm³ = 1 dm³

Step 2: Work out how many moles of the "unknown" stuff you must have had using the equation.

The equation's the same as above — one HCl for one NaOH, so you must have equal numbers of moles (of HCl and NaOH). So you must have 0.025 moles of HCl.

Step 3: Find the concentration of the "unknown" substance (i.e. the acid) using the same formula triangle as used above.

Concentration = number of moles ÷ volume
= 0.025 ÷ (30 / 1000) dm³
= 0.833 mol/dm³.

Check that your answers are sensible. In this case, you needed a bit more acid than alkali, so you'd expect the acid to be a bit less concentrated.

Concentrate on your calculations...

You need to be pretty careful with these calculations — always check that your answer is realistic. If you're given the volume of anything in cm³ you need to convert it to dm³ (because that's what the equation uses). It's pretty easy really — just divide the number by 1000 (see p.128 for more).

Investigating Properties of Materials

Materials scientists investigate the way substances behave in various conditions. They investigate the properties of materials, and use their results to suggest which materials are best for different jobs.

Materials Conduct Electricity Well If They Have Low Resistance

Say you're designing a robot to explore the surface of Mars. It'll need lots of electrical circuits to control its movement, cameras, data recording, etc. The various components (bits) of the circuits will need to be made of different materials depending on what they do.

1) Some materials conduct (carry) electricity better than others. E.g. copper conducts electricity well (that's why it's used for wiring in household circuits).

2) A 'good conductor' is a material that doesn't offer much resistance to the flow of electrical current.

3) Different substances have different electrical resistances. A materials scientist could find out the resistance of a new material or component using a test circuit.

You Can Use a Test Circuit to Work Out Resistance

With this circuit, you can measure the voltage across a component and the current flowing through it.

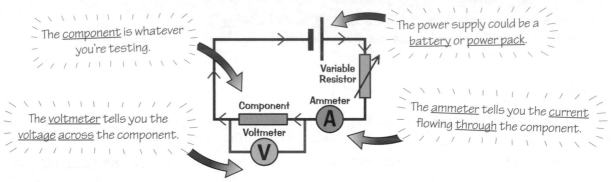

The component is whatever you're testing.

The power supply could be a battery or power pack.

Variable Resistor

Component

Ammeter

The voltmeter tells you the voltage across the component.

Voltmeter

The ammeter tells you the current flowing through the component.

Here's what you do:

1) Use the variable resistor to adjust the current to, say, 0.5 A.

2) Read the voltage from the voltmeter.

3) Record your current and voltage readings in a nice table. ➡

4) Use the variable resistor again to change the current to, say, 1.0 A. Record the new voltage across the component.

5) Keep on doing this until you've got several readings. (See the next page for why you should do this.)

Once you know the current and voltage, working out resistance is fairly easy. Use this equation:

Current (A)	Voltage (V)
0.5	2.4
1.0	

$$\text{Resistance} = \frac{\text{Voltage}}{\text{Current}}$$

EXAMPLE:

Current = 0.5 A, Voltage = 2.4 V.
So resistance = 2.4 ÷ 0.5 = 4.8 Ω.

This is the symbol for ohms — the ohm is the unit of resistance.

Test me — I won't resist...

The wire used in most electrical circuits has a very low resistance. If a fault develops in a circuit so that the only components are the power supply and some wire, it's called a short circuit. Short circuits can be dangerous — you can get a huge current (because of the low resistance), overheating, and a fire.

Investigating Properties of Materials

Sometimes it's useful if the <u>resistance</u> of a component <u>changes</u> as the <u>conditions change</u>. E.g. <u>electrical</u> <u>thermometers</u> work because the resistance of one of their components changes as the <u>temperature</u> varies.

Some Devices Have Constant Resistance...

Say you've tested a component (see the previous page), done the sums to work out resistance, and these are your results.

Current, I (in A)	1	2	3	4
Voltage, V (in V)	2.5	5.0	7.5	10.0
Resistance, R (in Ω)	2.5	2.5	2.5	2.5

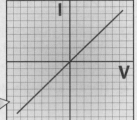

1) The <u>resistance</u>, R, is <u>constant</u> as the current increases.

2) If you plot current against voltage, you get a <u>straight line</u> through the origin — so you can see that the <u>current</u> is <u>proportional</u> to the <u>voltage</u>.

...And Some Have Variable Resistance

Not all components have a fixed resistance. For a <u>filament lamp</u>, the results would be a bit like this:

Current, I (in A)	1	2	3	4
Voltage, V (in V)	2.0	5.0	10.0	36.0
Resistance, R (in Ω)	2.0	2.5	3.3	9.0

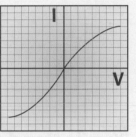

1) The resistance <u>isn't constant</u> — it increases as the current increases.
2) This is because the bigger the current through the <u>filament lamp</u>, the hotter it gets. And as its <u>temperature increases</u>, its <u>resistance increases</u>.
3) The graph's a <u>curve</u> — the <u>current</u> is <u>not proportional</u> to the <u>voltage</u>.

Resistance Depends on Length, Thickness and Material

A component in a circuit can be designed to have a certain resistance. Take a <u>piece of wire</u>, for example:

- When you build circuits in the lab, the wire you use is probably fairly <u>thin</u>, made of <u>copper</u>, and often comes in <u>ready-made lengths</u> (coated with black or red plastic insulation, with crocodile clips at either end).

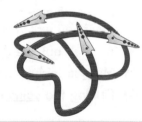

- If these wires were <u>thicker</u>, <u>longer</u> or made of <u>different material</u>, their <u>resistance</u> would be different. You might have to investigate how the resistance of a component is affected by its <u>length</u> and <u>thickness</u>, as well as <u>what it's made of</u>.

<u>EXAMPLE</u>: Investigating how <u>thickness</u> affects resistance.

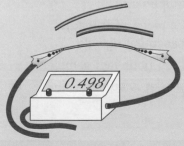

1) Use the standard test circuit, with a piece of wire as the 'component'.
2) Take several current and voltage readings (as on the previous page), then calculate the resistance for each one and work out the average (this gives you a much more reliable result than just doing it once).
3) Repeat step 2 with several pieces of wire of different thicknesses. All these pieces of wire must be the <u>same length</u> and made of the <u>same material</u> — so you can be sure it's the change in thickness which is causing any change in resistance.

- <u>Different materials</u> have <u>different resistances</u> — there's no hard and fast rule. For any one material, though, you should find that <u>thicker</u> wires have <u>lower</u> resistance but <u>longer</u> wires have <u>higher</u> resistance.

I'm not thick — I just have a low resistance...

With experiments like the one above, never just take one reading — take several (at least three). Then you'll spot whether the resistance is constant or variable — which is important. And if it's constant, you can use all those readings to take an <u>average</u> — just add them all up and divide by how many there are.

Investigating Properties of Materials

Materials scientists consider <u>physical properties</u> like density and strength as well as electrical properties like resistance. For instance, when you're building a <u>house</u>, choosing materials with the right physical properties is very important. E.g. we don't build houses from <u>chocolate</u> because it <u>melts</u> in warm weather.

Thermal Conductivity Means How Well Materials Conduct Heat

1) <u>Thermal conductivity</u> means how well a substance <u>conducts heat</u>. If the <u>walls of your house</u> had a <u>high</u> thermal conductivity, you'd be 'losing' a lot of heat to the air outside — which would be wasteful and expensive. It's better to build the walls of a house from materials with <u>low</u> thermal conductivity.

2) Here's a simple way to compare the <u>thermal conductivity</u> of <u>different materials</u>:

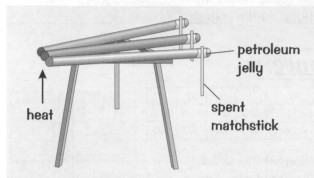

- Put three rods, all the <u>same size</u> but made of <u>different materials</u>, on a tripod, with a spent (used) matchstick stuck to one end with petroleum jelly.
- <u>Heat</u> the rods at the other end and time how long it takes each matchstick to fall off.
- The <u>sooner</u> a matchstick <u>falls off</u>, the quicker the heat has been conducted along the rod — so the <u>higher</u> the <u>thermal conductivity</u> of that material.

You'll Need to Compare the Densities of Different Materials

1) <u>Density</u> means how much <u>mass</u> there is in a certain <u>volume</u> of material. The more mass per cm³, say, the higher the density of the material — the more stuff is packed into the same space.

2) For example, imagine you're making a nice <u>cup of tea</u>. Now imagine you have two <u>teaspoons</u> exactly the same size and shape, but one's made of <u>steel</u> and the other one's made of <u>plastic</u>. The <u>steel</u> teaspoon would be a lot heavier — because steel is <u>denser</u> than plastic.

3) To work out the <u>density</u> of an object (like a teaspoon) you have to know its <u>mass</u> and its <u>volume</u>.

4) <u>Mass</u> is fairly easy to find — just use a <u>balance</u> to find the mass in grams (g).

5) Finding the <u>volume</u> of a funny-shaped object (like a teaspoon) is a bit harder. Here's how:

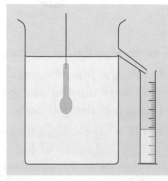

- Fill a <u>displacement can</u> with water (overfill it first and let water spill out — it will stop at the level shown opposite).
- Lower your test sample into the can (on a piece of thread). As you do this, water will spill into the measuring cylinder.
- This 'overspill' water has been 'displaced' by your test sample.
- The volume of <u>water</u> in the <u>measuring cylinder</u> is equal to the <u>volume of the test sample</u>. Clever.

6) Use this formula to calculate the density of the sample.
EXAMPLE: Mass = 14.92 g, Volume = 1.9 cm³.
So <u>density</u> = 14.92 ÷ 1.9 = <u>7.85 g/cm³</u>.

$$\text{Density (in g/cm}^3) = \frac{\text{Mass (in g)}}{\text{Volume (in cm}^3)}$$

Thermal conductivity — for pyjama-wearing orchestras...

You could find out your own density using the same method — weigh yourself to find your <u>mass</u>, then fill a bath to the top and get in (completely underwater) to find your <u>volume</u> — measure how much water spills over the top. You'll need to have a pretty big measuring cylinder though, to avoid floods.

Investigating Properties of Materials

Just as some people are stronger than others, some <u>materials</u> are stronger than others. And before building a tower block, say, it's a good idea to find some nice strong materials to build it with.

Testing Strength — Finding the 'Breaking Point'

1) There are different kinds of strength, e.g. how hard it is to <u>crush</u> something by <u>pressing down</u> on it, how hard it is to <u>snap</u> something by <u>stretching</u> it. Depending on what a material's being used for, one kind of strength will probably be more important.

2) E.g. the <u>wires</u> used to winch people onto <u>rescue helicopters</u> need to be strong in the sense of <u>not snapping</u> when heavy objects are dangled from them.

3) You could test materials for this kind of strength using the equipment shown below.

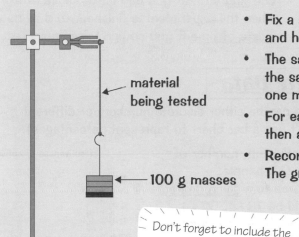

material being tested

100 g masses

- Fix a sample of each material to a clamp stand, as shown, and hang a <u>mass hanger</u> from the other end.

- The samples of each different material must all be exactly the same size and shape. E.g. if you're testing <u>wires</u>, each one must be the same <u>length</u> and the same <u>thickness</u>.

- For each sample, start off with the mass hanger empty, then add 100 g masses one by one until the sample breaks.

- Record the mass needed to break each sample. The greater that mass, the <u>stronger</u> the material.

Don't forget to include the mass of the mass hanger.

Material	A	B	C
Breaking mass (g)	700	700	900

You Could Make Your Results More Precise

From the results above, it looks like wire A and wire B have <u>exactly the same strength</u> — but this isn't necessarily the case:

1) You know that <u>700 g</u> was enough to break wire A, but it might have broken with a load of only 690 g, or 640 g, or 610 g — you can't tell from these results. It's the same thing for wire B.

2) So it <u>could be</u> that both materials are equally strong — but it <u>could be</u> that material A is stronger than B, or it could be that material A is <u>weaker</u> than material B. You'd have to do another experiment...

3) <u>Repeat</u> the experiment, but for wires A and B put <u>600 g</u> on the hanger straight away, then add <u>smaller masses</u> — just 10 g at a time, say, rather than 100 g.

4) To get a more precise result for <u>wire C</u>, you could put <u>800 g</u> on straight away, then add 10 g at a time.

5) The results from this second experiment might look something like this. These results are much more precise and you can now tell that material A is stronger than material B.

A	B	C
680 g	620 g	850 g

Working out strength — go to the gym...

When you're doing experiments, remember — <u>only change one thing at a time</u>. E.g. in the experiment above, each sample <u>must be</u> the <u>same length and thickness</u>. If you tested a <u>thin copper</u> wire and a <u>thicker bronze</u> wire, you wouldn't know what made the difference in strength — the type of <u>material</u> or the <u>thickness</u>.

Recording and Presenting Data

Well, by now you should have your standard procedure sorted and be more than familiar with the equipment that you'll be using. But it'll all be a waste of time if you don't record your results properly.

Always Use Tables to Record Results

The easiest way to record data during an experiment is by using a table.

1) Think about the data you're going to record and draw the table before you start — you should include columns for the data you are going to calculate (e.g. resistance) as well as the data you are going to measure (e.g. current and voltage).

2) Make sure you label your table clearly, showing what you are measuring and the units.

Length of wire (cm)	Voltage (volts, V)	Current (amps, I)	Resistance (ohms, Ω)
20			
40			
60			

It's important to keep your results table neat and tidy so that you can make sense of it when the experiment is finished. It'd all be a waste of time if you couldn't read your results.

Use Bar Charts to Present Discrete Data

1) Bar charts are used to present discrete data. This can be either discrete numbers or different categories, e.g. type of antibiotic (so you wouldn't use a bar chart to represent percentages).

Discrete data is data that can be measured exactly, e.g. number of people — you can't really have half a person or 2.2 people. Continuous data is data that can only be measured to a given degree of accuracy, e.g. a person's weight — this could be 76 kg, 52 kg, 64.2 kg. You'll never get an exact value because there'll always be a more accurate way of measuring, e.g. to the nearest g, mg or μg.

2) The width of the bars on a bar chart is always the same, but the heights of the bars vary to represent the data.

3) There should be a gap between each bar.

This chart shows the size of inhibition zones with different antiseptics.

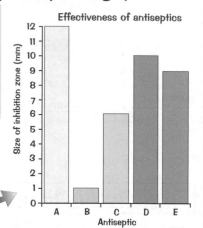

Use Histograms to Present Continuous Data

1) Histograms are used to present continuous data in different categories. They're tricky, but there's more about them in our GCSE Maths books (shameless plug).

2) The width of the bars can stay the same or vary to represent the size of the category.

3) The areas of the bars vary to represent the size of data.

Diameter of bacterial colony (mm)	Number of colonies	Frequency density of colonies
0-2	14	7
2-5	9	3
5-10	6	1.2
10-15	5	1
15-20	2	0.4

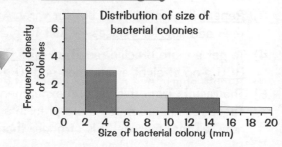

Pictograms can be used to present data in a visually appealing way.

1) They're bar charts or histograms where the bars have been replaced by pictures to represent the data.

2) The size or the number of pictures varies with the data.

3) You need to include a key to explain what the pictures mean.

Recording and Presenting Data

Use Pie Charts to Present Data as Proportions of a Whole

1) Pie charts are used to present data as <u>proportions</u> (<u>percentages</u> or <u>fractions</u>) of a whole. You'd never use them to represent raw data.

2) The 'pie' is divided into <u>segments</u> that represent proportions of the whole.

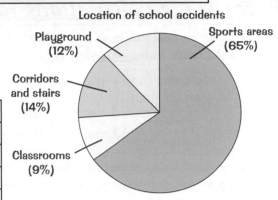

Location of school accidents

You can draw pie charts using a pie chart measurer or a protractor (though you'll need to convert each value into an angle).

Area of school	% of accidents
Sports areas	65
Playground	12
Corridors and stairs	14
Classrooms	9

Use Graphs to Show the Relationship Between Variables

1) You should use graphs when investigating the <u>effect</u> of <u>one variable</u> on <u>another</u>. They <u>shouldn't</u> be used when the data is in <u>different categories</u>. Note the difference between the following example (where the <u>same</u> antiseptic is being used but in <u>different concentrations</u>) and the bar chart on the previous page (where five <u>different</u> antiseptics were being used).

2) The variable that you <u>change</u> (e.g. the concentration of antiseptic used, or the length of wire used) is called the <u>independent variable</u>. This should be plotted on the <u>x-axis</u> (the horizontal axis).

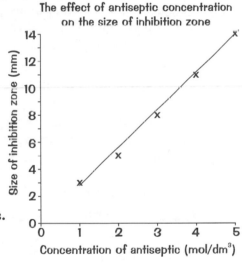

The effect of antiseptic concentration on the size of inhibition zone

3) The variable which is being <u>measured</u> (e.g. the size of the inhibition zone or the resistance of a piece of wire) is called the <u>dependent variable</u> and is plotted on the <u>y-axis</u> (the vertical axis).

4) Points on a graph can be <u>joined up</u> to make straight or curved lines. Don't forget that any line you draw is only really an <u>educated guess</u> — the data could do anything between the two points you measured.

5) The shape of the line shows the <u>relationship</u> between the variables.

6) Often a '<u>line of best fit</u>' is drawn — which goes through (or close to) most of the points on the graph.

7) Make sure your graphs have titles, their axes are labelled, and the scales are appropriate.

Other Visual Images can also be Useful

If it's appropriate, data can be represented in other ways — here are some examples:

1) <u>Radar charts</u> are a 'clock-face' form of bar chart — the bars <u>radiate</u> from a central point.

2) <u>Bubble charts</u> use <u>circles</u> of <u>different sizes</u> to display data.

3) <u>Cartograms</u> have <u>graphs</u> or <u>charts</u> placed on a <u>map</u> — they show how data <u>relates</u> to an area.

4) <u>Combination charts</u> — <u>two charts</u> in <u>one</u>, e.g. a <u>bar chart</u> and a <u>line graph</u> placed on the <u>same axes</u>.

5) <u>Diagrams</u> and <u>sketches</u> can be useful — remember to <u>label</u> drawings and include a <u>scale</u>.

6) <u>Photographs</u> or <u>video footage</u> can also be used to display results in a very <u>visually appealing way</u>.

Pie chart — a new entry at number two: meat and potato...

Sorry to go on, but I just can't stress this enough — don't forget to <u>label your axes</u> and add a <u>title</u>.

Calculations and Analysis

Sometimes you need to <u>convert</u> your results into different <u>units</u> so you can use them in <u>formulas</u>.

For Some Calculations You Need to Change the Units

Here's an example of how you change the units for <u>mass</u> (formulas with mass in normally need the units to be <u>kilograms</u> (kg)).

| microgram (µg) |
| milligram (mg) |
| gram (g) |
| kilogram (kg) |

There are <u>1000 µg</u> in a <u>milligram</u>, so to convert 1 microgram into mg you ÷1000.

There are <u>1000 mg</u> in a <u>gram</u>, so to convert 1 milligram into grams you ÷1000.

There are <u>1000 g</u> in a <u>kilogram</u>, so to convert 1 gram into kilograms you ÷1000.

Similar rules apply if you're going the <u>other way</u>. You just <u>multiply</u> instead of <u>divide</u>. E.g. if you have 5.2 kg of salt and want to know how many grams that is, you'd multiply by 1000: 5.2 × 1000 = 5200 g.

The words <u>before</u> the unit tell you how big it is. E.g. <u>kilo</u> means that there's <u>1000</u> of the unit (<u>kilogram</u> = 1000 grams). <u>Milli</u> means <u>one thousandth</u> (<u>milligram</u> = one thousandth of a gram).

So, if you want to convert a SMALL unit to a BIGGER one you DIVIDE. If you want to convert a BIG unit to a SMALLER one you MULTIPLY.

You Need to Know These Units

1) <u>Volume</u> — Most formulas use <u>cubic decimetres</u>. There are 1000 <u>cubic decimetres</u> (dm³) in one <u>cubic metre</u> (m³), 1000 <u>cubic centimetres</u> (cm³) in a cubic decimetre and 1000 <u>cubic millimetres</u> (mm³) in a cubic centimetre.

1 dm³ is the same as 1 litre. 1 cm³ is the same as a millilitre (ml).

2) <u>Length</u> — Most formulas use <u>metres</u>. There are 1000 <u>metres</u> (m) in one <u>kilometre</u> (km), 100 <u>centimetres</u> (cm) in a metre, 10 <u>millimetres</u> (mm) in a centimetre and 1000 <u>micrometres</u> (µm) in a millimetre.

3) <u>Time</u> — Most formulas use <u>seconds</u> (there are <u>60</u> seconds in a minute). To convert <u>minutes</u> (min) to seconds (s) you need to <u>multiply by 60</u>. To convert <u>hours</u> (h) into minutes you also <u>multiply by 60</u>.

4) <u>Temperature</u> is measured in <u>degrees Celsius</u> (°C). (You won't need to convert it into any other form.)

Here are some physics units you need to know:

1) <u>Current</u> — measured in <u>amperes</u> (A), there are 1000 <u>milliamperes</u> (mA) in one ampere.

2) <u>Resistance</u> — measured in <u>ohms</u> (Ω), there are 1000 ohms in a <u>kilohm</u> (kΩ) and 1000 kΩ in a <u>megohm</u> (MΩ).

3) <u>Energy</u> — measured in <u>joules</u> (J), there are 1000 J in one <u>kilojoule</u> (kJ).

4) <u>Power</u> — measured in <u>watts</u> (W), there are 1000 W in one <u>kilowatt</u> (kW).

5) <u>Density</u> — measured in <u>grams per cubic centimetre</u> (g/cm³) or <u>kilograms per cubic metre</u> (kg/m³).

6) <u>Velocity</u> — measured in <u>metres per second</u> (m/s). Acceleration is <u>metres per second squared</u> (m/s²).

7) <u>Voltage</u> (or potential difference) — measured in <u>volts</u> (V).

8) <u>Force</u> — measured in <u>newtons</u> (N).

There aren't as many units used in chemistry, but you still need to learn a few:

1) <u>Chemical quantity</u> — measured in <u>moles</u> (mol).

2) <u>Concentration</u> — measured in <u>grams per cubic decimetre</u> (g/dm³), which is the same as <u>grams per litre</u> (g/l). Concentration can also be measured in <u>moles per cubic decimetre</u> (mol/dm³).

My friend's really rich — he lives in a megohm...

Remember, if there's a <u>kilo before the unit</u>, then there's <u>1000</u> of that unit. Easy peasy.

Calculations and Analysis

It's all very well collecting all your data and making it look nice with pretty little graphs, but you also need to be able to interpret your results and spot if any of them look a bit odd.

Repeat Experiments and Take an Average of the Results

You should repeat experiments at least three times, and take an average of the results
— this improves reliability and helps you spot any 'anomalous' results (i.e. results that don't seem right).

1) Look for any anomalous results.
2) Repeat experiments that produced anomalous results.
3) Add up the consistent readings.
4) Divide the total by the number of readings to give the average.

Investigating the effect of wire length on resistance

Repeat	Length of wire (cm)	Voltage (V)	Current (A)
1	10.0	0.60	0.95
2	10.0	0.60	0.94
3	10.0	0.60	0.76
4	10.0	0.60	0.92
		Average	0.94

EXAMPLE — working out the average resistance:

1) The third repeat gave a much lower result than the others — this result is anomalous.
2) The experiment has been repeated (repeat 4).
3) Add up repeats 1, 2 and 4: (0.95 + 0.94 + 0.92 = 2.81).
4) Calculate the average: 2.81 ÷ 3 = 0.94 A

Analyse Your Results Using Calculations, Graphs and Charts

Once you have finished the experiment and calculated the average of your results, you may need to use calculations to analyse your results. This means using formulas to convert data into more useful forms.

Length of wire (cm)	Voltage (V)	Average current (A)	Resistance (Ω)
10.0	0.60	0.94	0.64
20.0	0.60	0.43	1.40
30.0	0.60	0.29	2.07
40.0	0.60	0.20	3.00

EXAMPLE — work out resistance from readings of current and voltage using the formula $V = I \times R$.
You need to rearrange the formula to get resistance $(R = V \div I)$. For the 10 cm wire:
$R = 0.6 \div 0.94 = 0.64\ \Omega$

Once your results are in a useful format you might want to plot them on a graph or chart. You'll need to think about: 1) The type of graph or chart you choose to display your results.
2) A suitable scale. Make sure you plot all points correctly.
3) Whether or not a line of best fit is appropriate.
4) Any anomalies (results that don't look right), and trends or patterns in the results.

EXAMPLE

1) A graph has been drawn because we're investigating the relationship between two variables (i.e. the effect of wire length on resistance).
2) The maximum resistance is 3 Ω, so increments of 0.5 Ω gave a suitable scale (anything bigger would've made the graph too small).
3) A line of best fit was drawn to show the trend.
4) As the length of the wire was increased the resistance also increased.

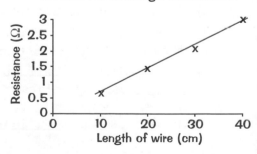
The effect of wire length on resistance

None of my results are anomalous — they all have names...

Getting your chart or graph right is crucial. But, first you need to decide which type to draw. Remember, if you're investigating the relationship between variables then you should draw a graph (see p.127 for more).

Conclusions and Evaluations

Planning an experiment isn't too bad and carrying it out can be quite fun — but the worst part, without a shadow of a doubt, is evaluating — picking holes in your own work just isn't natural, but it has to be done.

Conclusions Should Summarise and Explain Results

At the end of your investigation you need to include a <u>conclusion</u>. You should:

1) <u>State</u> what your <u>results show</u>.

2) <u>Describe</u> your <u>graphs or charts</u>.

3) <u>Identify</u> any <u>patterns or trends</u>.

4) <u>Explain</u> your conclusions using <u>science</u> — the more <u>understanding</u> you show the better.

EXAMPLE

1) The results show that the <u>longer</u> the piece of wire the <u>greater</u> the resistance.

2) The graph has a <u>straight line</u>, which means that the resistance and the length of the wire have a linear relationship.

3) The graph shows that <u>resistance increases</u> by approximately 0.75 Ω for <u>every 10 cm of wire</u>.

4) Resistance is a measure of how hard it is for current to flow around a circuit. The longer the wire, the harder it is for current to flow. So the longer the wire, the higher the resistance.

Evaluations — Describe How You Could Improve It

1) Comment on your <u>method</u> — was the <u>equipment suitable</u>? Did the <u>procedure</u> allow you to obtain <u>accurate results</u>?

2) Comment on the <u>quality of your results</u> — did you get <u>enough evidence</u> to reach a <u>conclusion</u>? Were your results <u>reliable</u>?

3) Identify any <u>anomalies</u> in your results — if there were <u>none</u> then <u>say so</u>.

4) Try to <u>explain</u> any anomalies — were they caused by <u>errors in measurement</u>? Were there any other <u>variables</u> that could have <u>affected your results</u>?

5) Suggest any <u>changes</u> that would <u>improve your investigation</u> — is there more <u>suitable equipment</u> you could have used? What further work could provide <u>additional evidence</u> to support your conclusions?

EXAMPLE

1) The equipment and method used were suitable and produced accurate results. But the <u>degree of accuracy</u> might have been affected by the <u>voltmeter</u> and <u>ammeter</u> (if they weren't <u>calibrated</u> properly). There may also have been a slight variation in the thickness of the wire.

2) The results were reliable. Each experiment was <u>repeated</u> and calculations were carried out on the <u>average</u> readings for each length. The number of readings taken was enough to reach a valid <u>conclusion</u>.

3) One of the results for 10 cm of wire was <u>anomalous</u>. This experiment had to be <u>repeated</u>.

4) The anomalous result could have been caused by an <u>error in the measurement</u> of the length of the wire, a difference in the <u>thickness</u> of the wire, or an error when taking <u>readings from the ammeter or voltmeter</u>.

5) If the experiment was carried out again then <u>no changes</u> would be made to the <u>equipment</u> or <u>procedure</u> used, but extra care would be taken when measuring the lengths of wires. Other experiments could be carried out, e.g. testing <u>more lengths</u> or keeping the length of wire the same but <u>changing the thickness</u>.

I like experimentation — draw your own conclusions...

And that's pretty much that. By now you should know <u>everything</u> there is to know about <u>carrying out a scientific investigation</u>. I hope you've been paying attention because you're going to have to do it yourself. If you're still a bit <u>uncertain</u> there's more <u>help</u> on writing up an investigation on page 144.

Report: Developing Scientific Skills

It's all very well being a genius with a burette, but you'll get no marks unless you <u>write up</u> your experiments.

You Need to Write Three Reports About Practical Activities

You need to write THREE reports about your experiments — one for each of these areas:
1) Living Organisms, 2) Chemical Analysis, 3) Properties of Materials.

These reports will make up your portfolio for <u>UNIT 3: DEVELOPING SCIENTIFIC SKILLS</u>.

<u>Each report</u> will have <u>two bits</u> to it:

1) A <u>WRITE-UP</u> of <u>all</u> the activities (there are several for each report). For each activity, include:
   - a <u>risk assessment</u> and a <u>description of what you did</u>,
   - your <u>results</u>,
   - <u>analysis</u> of your results and a <u>conclusion</u>,
   - an <u>evaluation</u> of the activity.
2) A <u>DESCRIPTION</u> of how each practical activity is <u>used in a workplace</u>.

It's important to evaluate your activity.

Include a Risk Assessment and Describe What You Did

1) Before you start each activity, <u>read the instructions</u> for it and do a <u>risk assessment</u> (see p.7).
2) Choose the <u>right equipment</u> and <u>set it up carefully</u> (see p.109-p.113).
3) Write a <u>list</u> of all the equipment you're going to use, and draw <u>labelled diagrams</u> where appropriate.

Make Measurements and Present Your Results Carefully

1) Do the activity, following any instructions step by step. You're marked on the <u>accuracy</u> of your observations, so you need to take measurements carefully and record your results in tables, etc.
2) Remember to <u>repeat</u> observations and measurements where necessary:
   - repeating tests and working out an <u>average</u> will give you a <u>more reliable result</u>,
   - if a result seems obviously <u>wrong</u>, do the test again (and write down why you're repeating it).
3) <u>Present your results</u> sensibly — tables, charts, graphs, etc. (See p.126-p.127 for help with these.)

Identify Relationships in Your Conclusion

1) Your data might not be very <u>useful</u> in its 'raw' form — so you might have to do some <u>calculations</u>, e.g. to work out resistance from measurements of voltage and current.
2) Look for <u>patterns</u> in your results, e.g. straight lines on graphs, and say what they mean (see p.129).
3) Draw an overall <u>conclusion</u> — say what the <u>evidence</u> from your experiment has shown (see p.130).

Evaluate the Activity and Describe a Workplace Application

1) <u>Evaluate</u> what you've done, e.g. how <u>reliable</u> do you think your data is? How could you make it <u>better</u>?
2) Describe <u>what kinds of organisation</u> might use the activity, and <u>why</u>.

Develop your scientific skills — try juggling with burettes...*

Practical activities are nothing to be scared of — often, it's just a case of <u>following instructions</u> and <u>taking measurements carefully</u> — then <u>recording everything clearly</u> (see p.126).   *Actually, no, don't.

Report: Monitoring Living Organisms

Scientists monitor organisms to help them understand how the organisms <u>grow</u>, <u>develop</u> and <u>behave</u>.

You Need to Write a Report on Monitoring a Living Organism

This will be the FIRST OF FOUR reports that make up your portfolio for
<u>UNIT 4: USING SCIENTIFIC SKILLS FOR THE BENEFIT OF SOCIETY</u>.

Your report will have <u>two bits</u> to it:

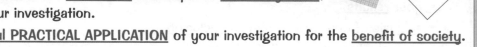

1) A <u>WRITE-UP of your investigation</u>, including:

 • A <u>plan</u> of your investigation,

 • Your <u>results</u>,

 • <u>Analysis</u> of your results and a <u>conclusion</u> that explains <u>what they mean</u>,

 • An <u>evaluation</u> of your investigation.

2) A <u>description</u> of a <u>useful PRACTICAL APPLICATION</u> of your investigation for the <u>benefit of society</u>.

Plan Your Investigation Carefully

You can't do an investigation without a plan, and a <u>good plan</u> can make all
the difference to your marks. Before you start, you'll need to decide:

1) What type of <u>organism</u> you're going to investigate, and what
the <u>purpose</u> of your investigation is — <u>what</u> to investigate and <u>why</u>.

2) How to look after the <u>welfare</u> of the organisms you're investigating,
and minimise <u>risks</u> to yourself (see next page).

3) What <u>conditions</u> you'll be controlling and how you'll make it a <u>fair test</u>.

4) What equipment you'll use to <u>monitor</u> the organism, <u>how often</u>
you'll monitor it, and for <u>how long</u>.

This observation hadn't
been in Jeremiah's plan

1) Choose Your Organism and Decide What to Investigate

You'll need to <u>CHOOSE</u> what you want to investigate — you could try investigating:

1) How <u>changing the conditions</u> affects the <u>yield of product</u> from a plant or microorganism.

2) The effects of <u>physical activity</u> on <u>human beings</u>.

3) How <u>changing the conditions</u> affects the <u>growth</u>, <u>behaviour</u> or <u>development</u> of an organism.

In theory you could investigate <u>any organism</u> you like, but you have to be <u>realistic</u> (so lions
and wildebeest are probably out). Once you've got an idea for an investigation, ask yourself:

1) Will it fit in with the <u>TIME</u> that's available? If you've only got a couple of weeks for the task,
then anything involving the growth of plants is unlikely to fit in.

2) Does my school or college have all the <u>EQUIPMENT</u> I need to carry out my investigation?
You might need to speak to the technicians, so they can order anything that isn't in stock.

3) Is the <u>COST</u> reasonable? Nobody likes getting turned down, or wasting their time, so make
sure you find out if something's too expensive before planning too much of your investigation.

4) Can I take appropriate <u>CARE</u> of any living organisms that I'm planning to use?

And I had my heart set on a yeti — you win some, you lose some...

These three pages should give you a few ideas about the <u>sort of investigation</u> you could do (as well as
what you <u>can't</u> do) and some advice on what you'll need to put in your report. Read and enjoy...

Report: Monitoring Living Organisms

2) Think About the Welfare of You and Your Organisms

If you're carrying out an experiment on a living organism, you have to make sure that any <u>discomfort</u> or <u>distress</u> caused is kept to an <u>absolute minimum</u>.

In practice, that means you're <u>very limited</u> in your choice of which animals you can use — in fact, you'll probably be restricted to <u>humans</u>. With a human subject, you have to carry out a risk assessment (see p.6-11) to make sure your experiment's <u>safe</u>, and that both you and your subject know about <u>any potential risks</u>.

The welfare of <u>plants</u> and <u>microorganisms</u> is less of an issue, which makes them <u>much easier</u> to investigate than animals, but you still need to do a <u>risk assessment</u> for your <u>own safety</u> (and marks).

3) Decide What Conditions You'll Vary and What You'll Keep Constant

<u>Whatever</u> your investigation, it will involve <u>varying</u> something and monitoring the effect on your organism. At the same time you'll need to try your best to keep <u>all the other factors constant</u> so that they don't affect the results of your investigation.

Which conditions you vary depends on <u>what you want to find out</u>.

1) If you're looking at the <u>yield of products</u> from, or <u>growth</u> of, plants or microorganisms, you could vary the <u>temperature</u>, the amount of <u>nutrients</u>, the amount of <u>water</u>, or <u>light</u> levels.

> <u>EXAMPLE</u>: Plants need minerals to grow well, e.g. nitrates, phosphates and potassium.
> You could investigate the effect of varying the concentration of one of these minerals. You would need to supply different groups of plants with feeds containing different amounts of the mineral. To make it a fair test, you would have to keep all the other conditions (e.g. light, temperature, water) the same for each group.

2) If you're investigating the effects of <u>physical activity</u> on <u>human beings</u> then you might think about changing either the intensity or the duration of a period of exercise.

4) Decide How, How Often and for How Long to Monitor the Organism

If you're investigating the effects of <u>physical activity</u> on humans, you could measure:

1) <u>Heart rate</u> — by counting the number of <u>pulses</u> in your subject's wrist.

2) <u>Breathing rate</u> — by counting the number of breaths your subject takes.

You'll only need to take measurements over <u>a minute</u> or so, <u>before</u> and <u>after</u> exercise.

If you're investigating growth or yield of product using <u>plants</u>, you could measure:

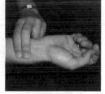

taking a pulse

1) <u>Growth</u> — <u>height</u> of seedlings or <u>number</u> of leaves.

2) <u>Yield</u> — <u>mass</u> of vegetable or fruit produced.

Plants tend to grow quite <u>slowly</u>, but the rate of growth (and time taken to produce vegetables or fruits) <u>varies</u> from <u>species to species</u>. With a fast-growing plant, you may need to take measurements every <u>two or three days</u> for a period of <u>four</u> to <u>six</u> weeks (depending on the plant).

To find growth or yield of product using <u>microorganisms</u>, you could measure:

1) <u>Growth</u> — how the <u>size</u> or <u>number</u> of colonies changes over time.

2) <u>Yield</u> — <u>how much</u> product, e.g. alcohol, is produced in a given time.

Because microorganisms reproduce very quickly you might find you need to take measurements <u>every day</u> or <u>every other day</u> for a <u>week</u> or a <u>fortnight</u>.

> <u>EXAMPLE (cont)</u>: A good plant to use would be Brassica rapa — 'rapid cycling brassica'. Plants can be grown from seed in about 4 weeks — growing to around 25 cm tall. You could measure the heights of the seedlings every 2-3 days.

Report: Monitoring Living Organisms

Collect Relevant Data and Record It Clearly

1) Your plan should say clearly <u>what</u> you're going to measure and how you're going to measure it. You'll need to take <u>repeat measurements</u>, and work out <u>averages</u>, to make your results more <u>reliable</u>.

2) So if you're growing plants or microorganisms, you need <u>more than one</u> plant or culture plate for <u>each condition</u>. If you're investigating the effect of activity on a human you'll need to get them to <u>repeat the activity</u> several times so you can take repeat measurements — lucky subject.

3) <u>Check</u> any results that look weird by repeating part of the investigation.

4) The best way to present your results is to use <u>tables</u> and <u>graphs</u> to show any <u>trends</u> in the data.

<u>EXAMPLE (cont)</u>: If you were measuring the growth of *Brassica rapa* seedlings at different concentrations of nitrate, you might record your data like this:

Concentration of nitrate (mg/l)	Height of seedlings after 7 days (mm)			
	Seedling 1	Seedling 2	Seedling 3	Average
0.1	51	49	50	50
0.2	57	54	55	55.3

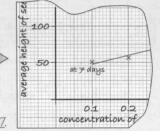

You might want to gather the same sort of data at, e.g., 14 days, 21 days and 28 days, and plot them separately on the graph. <u>For help with tables and graphs, see pages 126-127.</u>

Analyse Your Results and Evaluate Your Investigation

<u>ANALYSING</u> your results (see p.129) means you have to:

Look for <u>trends</u> and <u>patterns</u> in your data. Make <u>conclusions</u>, based on the trends, to <u>explain</u> your results. Comment on any data that <u>doesn't fit</u> the pattern — how <u>reliable</u> are your results?

<u>EVALUATING</u> your investigation (see p.130) means you have to think about:

What went <u>well</u>, what <u>didn't</u> go so well and <u>why</u>. How you might <u>improve</u> the investigation in the future. Whether there's anything you could do <u>differently</u> to get more <u>reliable results</u>.

And Finally... Describe a Useful Application of Your Investigation

<u>MONITORING THE HUMAN BODY</u> is essential in the <u>diagnosis</u> and <u>treatment</u> of disease. Measurements of <u>breathing rate</u> and <u>heart rate</u> are used as part of general health checks, as well to help <u>athletes</u> monitor their fitness for training. Other measurements, like blood sugar levels and blood pressure, are used to monitor specific disorders.

<u>MONITORING THE GROWTH AND DEVELOPMENT OF PLANTS</u> is very important in <u>agriculture</u>. Agricultural scientists work very closely with <u>crop farmers</u> and <u>nurseries</u> to work out what conditions plants need for <u>healthy growth</u> and <u>maximum yield</u>.

<u>MONITORING MICROORGANISMS</u> has lots of applications — from developing <u>new drugs</u> to <u>brewing beer</u>. Some microbiologists work with <u>genetically modified</u> bacteria to produce <u>useful products</u>, e.g. human insulin. They have to make sure the conditions are just right to produce the <u>maximum yield</u>.

<u>EXAMPLE (the end)</u>: Farmers want to get quick, healthy growth and high yields from their crops. Agricultural scientists use experiments like this one to find the perfect combination of nutrients for each type of crop. The end result is a high yield of good quality crops for the farmer to sell and plenty of cheap food for the consumer.

Babysitting — monitoring microorganisms...

When you record your data, use <u>neat tables</u> right from the start. Don't just scribble something down on a paper towel to write up later — that's how you lose results and make mistakes. Okay, nag over.

Report: Making a Useful Product

Industrial chemists use many different types of reaction to make products. Chemical industries try to maximise profits by making as much product as possible from the starting materials — a high yield.

You Need to Write a Report on Manufacturing a Chemical

This will be the SECOND report that goes into your portfolio for
UNIT 4: USING SCIENTIFIC SKILLS FOR THE BENEFIT OF SOCIETY.

You'll be given a PROCEDURE to follow to make a product — your report will be in three bits:

1) An explanation of the UNDERLYING CHEMISTRY involved in the reaction:
   - Identify the type of reaction used.
   - Write a balanced chemical equation to describe the reaction.
   - Describe and explain the factors affecting the rate of the reaction.

2) Your PREPARATION of a pure, dry product, including:
   - The product itself, presented in a suitable sample tube, labelled with its name, date of preparation and correct hazard symbols (see p.6).
   - A measurement of the actual yield of the product.
   - A calculation of the percentage yield of your product.
   - A calculation of the costs of making a certain amount of the product.
   You'll also need to include a RISK ASSESSMENT.

3) An explanation of the INDUSTRIAL IMPORTANCE of the reaction and your product.

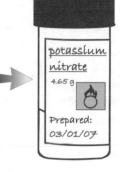

potassium
nitrate
4.65 g

Prepared:
03/01/07

Identify the Type of Reaction Used to Make Your Product

There are five main types of chemical reaction used to manufacture products:

1) OXIDATION — a substance gains oxygen.
   E.g. when the ore lead sulfide is heated with oxygen, it's oxidised to make lead oxide.

2) REDUCTION — a substance loses oxygen.
   E.g. when lead oxide is heated with powdered carbon, it's reduced to make solid lead. *Reduction in a furnace*

3) NEUTRALISATION — an acid and an alkali react together to produce a neutral compound called a salt.
   E.g. when potassium hydroxide neutralises nitric acid, potassium nitrate (a salt) and water are formed.

4) PRECIPITATION — an insoluble solid is formed after two solutions are mixed.
   E.g. mix solutions of lead nitrate and potassium chromate, and you get a precipitate of lead chromate.

5) THERMAL DECOMPOSITION — heat energy is used to break up a compound into simpler substances.
   E.g. when limestone is heated it breaks down into quicklime (calcium oxide) and carbon dioxide.

Write a Balanced Symbol Equation to Represent the Reaction

Chemical engineers use balanced symbol equations to help them decide in what proportions they need to mix the starting materials. You'll need to write an equation for your reaction — do it step by step:

1) Start with the word equation.   EXAMPLE: potassium hydroxide + nitric acid → potassium nitrate + water

2) Work out or look up the symbols for your chemicals, and replace each of the words in your equation with its symbol.   EXAMPLE (cont): $KOH + HNO_3 \rightarrow KNO_3 + H_2O$

3) Count up the number of each type of atom on each side of the equation — if you get different numbers, you'll have to balance your equation (see p.68). This one's already balanced.

You'll need to write a bit about the rate of reaction, too — but I'll come back to that.

Report: Making a Useful Product

Your teacher will give you a procedure to follow to actually make your product. You'll need to do a <u>full</u> <u>RISK ASSESSMENT</u> (see p.7) before you start work — and <u>keep a record</u> of it to put in your report.

> <u>EXAMPLE (cont)</u>: KOH is a strong alkali and HNO₃ is a strong acid. Both are <u>harmful</u> and <u>corrosive</u>.
> When handling either chemical, always wear safety goggles and take care to avoid contact with the skin.
> The product is <u>oxidising</u>. Avoid contact between the product and combustible materials (e.g. wood or paper).

Present Your Product Purified...

<u>Purification</u> in your case will probably mean separating your <u>solid product</u> from a <u>liquid</u>.

There are <u>two ways</u> to do that, depending on whether or not your product is <u>soluble</u> (see p.92).

1) <u>FILTRATION</u> — used to separate an <u>insoluble solid from a liquid</u>.

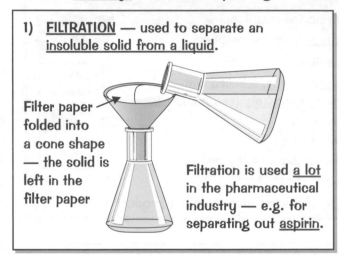

Filter paper folded into a cone shape — the solid is left in the filter paper

Filtration is used <u>a lot</u> in the pharmaceutical industry — e.g. for separating out <u>aspirin</u>.

2) <u>EVAPORATION</u> — used to crystallise a <u>soluble solid from solution</u>.

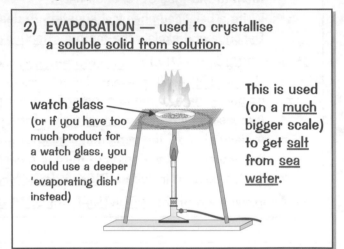

watch glass (or if you have too much product for a watch glass, you could use a deeper 'evaporating dish' instead)

This is used (on a <u>much</u> bigger scale) to get <u>salt</u> from <u>sea water</u>.

> <u>EXAMPLE (cont)</u>: KNO₃ is soluble in water, so is purified by evaporation.

...and Dried

Once you've separated out your solid, you need to <u>dry it</u>.

There are <u>three ways</u> of drying a solid product. Which you use depends on your <u>product</u> and your <u>timing</u>:

1) Leave it to <u>dry in the air</u> — this is the <u>easiest</u> and <u>cheapest</u> method. You get <u>better results</u> if there's a <u>dehumidifier</u> in the room, but even so, it might take <u>several days</u> to get a dry sample.

2) If you're in a hurry you could use a <u>drying oven</u>. Some drying ovens work just like an oven in a kitchen — they simply <u>heat</u> the sample to dry it.

You can also get ones that are more like <u>hairdriers</u> — they blow a stream of hot, dry air through the powdered sample. These are used a lot in the <u>pharmaceutical industry</u>, because they stop the drug clumping up.

GEOFF TOMPKINSON / SCIENCE PHOTO LIBRARY

3) Or (on the off-chance your school has one) you could use a type of drier called a <u>desiccator</u>, which doesn't use heat. You'd only need to use one for something that breaks up on heating.

> You'll need to <u>weigh</u> your sample at intervals during the drying process (how often you need to weigh it depends on the method you're using to dry your sample). If the mass <u>stays the same</u> from one weighing to the next, your sample's <u>dry</u>. You'll need to make a note of the dry mass of your product for later — more about that on the next page.

Chemistry can be dull — like watching potassium nitrate dry...

It's really important to carry out a risk assessment <u>before</u> you start doing anything — especially in a chemistry lab. Always read the labels on the chemical bottles, wear goggles, tie long hair back, etc...

Report: Making a Useful Product

In industry it's <u>not enough</u> to just make a bit of product and whack it in a tub. You need to know exactly <u>how much</u> product you've made, how <u>efficient</u> your reaction is, and how much each gram of product <u>costs</u> to make. It can get pretty technical and there are <u>calculations</u> to do. Great.

Find the Actual Yield and Percentage Yield of Your Product

You need to understand the <u>difference</u> between the <u>actual yield</u>, the <u>theoretical yield</u> and the <u>percentage yield</u> of your product:

1) <u>ACTUAL YIELD</u> — this is the <u>mass</u> of <u>pure, dry product</u> that you end up with. It depends on the amount of reactants you started with. To find the actual yield of your dried product:
   • <u>Weigh</u> the watch glass, evaporating dish or filter paper <u>with your product in</u> — call that mass 1.
   • <u>Transfer</u> your product to a sample tube, then <u>reweigh</u> the <u>empty</u> glass, dish or paper — mass 2.
   • <u>Subtract</u> mass 2 from mass 1 to give the <u>actual yield of product</u> that you'll need for your report.

2) <u>THEORETICAL YIELD</u> — this is the <u>maximum possible mass</u> of pure product that <u>could</u> have been made using the amounts of reactants you started with. It's calculated from the balanced symbol equation.

3) <u>PERCENTAGE YIELD</u> — this is the <u>actual yield</u> of the product as a <u>percentage</u> of the <u>theoretical yield</u>.

$$\text{Percentage Yield} = \frac{\text{Actual Yield}}{\text{Theoretical Yield}} \times 100\%$$

The percentage yield will <u>always be less than 100%</u>. That's because (among other reasons) some product will be lost along the way, e.g. during purification, drying and moving between containers. If your yield comes out at <u>greater than 100%</u> then your product probably isn't properly dry.

> <u>EXAMPLE:</u> 100 cm³ of 1 mol/dm³ KOH solution reacted with 1 mol/dm³ HNO_3 to give an <u>actual yield</u> of <u>4.65 g</u> of KNO_3.
> (cont) The <u>theoretical yield</u> for this reaction was <u>10.1 g</u>.
> So the <u>percentage yield</u> was (4.65 / 10.1) × 100% = <u>46%</u>

For <u>top marks</u> you might need to calculate the theoretical yield yourself. Ask your teacher about it.

Work Out the Cost of Making Your Product

1) Find out the <u>cost</u> of <u>each reactant</u> from your teacher or lab technician. You will probably be given prices <u>per kg</u>, or <u>per dm³</u> for solutions (of a particular concentration in mol/dm³ — see p.119).
2) Find the costs <u>per gram</u> or <u>per cm³</u> for each reactant.
3) For solid reactants, multiply the cost per gram by the number of grams <u>you used</u> for each reactant. Similarly for solutions — multiply the cost per cm³ by the volume you used (in cm³).
4) Then to find the <u>total cost</u>, add together the costs of the <u>individual reactants</u>.

> <u>EXAMPLE:</u> 100 cm³ of 1 mol/dm³ KOH solution reacted with 100 cm³ of 1 mol/dm³ HNO_3 to make 4.65 g of KNO_3.
> (it's still going)
> 1) Cost of the KOH = 27p per dm³ 1) Cost of the HNO_3 = 94p per dm³
> 2) Cost per cm³ of the KOH = 0.027p per cm³ 2) Cost per cm³ of the HNO_3 = 0.094p per cm³
> 3) Cost of KOH used = 0.027 × 100 = <u>2.7p</u> 3) Cost of HNO_3 used = 0.094 × 100 = <u>9.4p</u>
> So, the total cost to make 4.65 g of KNO_3 is 2.7p + 9.4p = <u>12.1p</u>

Aaaaaarrrrghhhh — run away...

In industry there'll be other costs as well — <u>research</u>, <u>insurance</u>, <u>wages</u>, <u>equipment</u>, <u>energy</u> bills, etc.

Report: Making a Useful Product

And now for a bit more underlying chemistry...

Five Factors Can Affect the Rate of Your Reaction

The rate of a reaction is the speed at which it happens. In industry it's really important that reactions happen as fast as possible (as long as they're controllable) — since the faster the product is made the more profitable the process.

The rate of a reaction can be increased in FIVE WAYS:

1) By increasing the temperature of the reaction.
2) By breaking up solids into smaller pieces — so powder reacts faster than big lumps.
3) By using higher concentration solutions.
4) By increasing the pressure in a reaction between gases.
5) By using a catalyst (some reactions don't have a catalyst).

You'll need to describe how TWO of these factors affect the rate of your reaction.

For top marks, you also need to be able to say why these factors affect the rate.

For a reaction to happen, particles have to bump into each other with enough energy to react. The rate of a reaction increases if particles collide more often and with more energy.

1) Using a higher temperature makes particles move faster, so they collide more often. It also gives the reacting particles more energy.
2) In lumps of solid, most of the particles are trapped in the middle of a lump. Grinding it up into a fine powder increases the number of particles at the surface that can take part in collisions.
3) Solutions with a higher concentration contain more particles per cm³ of solution, so there will be more collisions in the reaction mixture.
4) At high pressure, gas particles are closer together, so again there will be more collisions.
5) A catalyst speeds up a reaction without being used up itself. Catalysts reduce the energy that particles need to have in order to react.

Describe How Your Product is Used

Now you've made your product, you have to give one use for it and describe how it's used for that purpose.

Most chemical products have more than one use, so don't pick something obscure that you don't know anything about.

For top marks, you also need to:

Explain why this use is important and describe its impact on society.

EXAMPLE (and now the end is here...):

Potassium nitrate has many uses, from a food preservative to an ingredient in fertilisers. One of its first uses was in gunpowder and fireworks.

Potassium nitrate is oxidising, so it helps the fuel in a firework or signal flare to burn quickly and brightly, so they can be seen for miles around.

The military uses of gunpowder have had a huge impact on society. Guns and cannons completely transformed warfare.

How to get a fast, furious reaction — prod your teacher...

Phew — things all got a bit hairy there towards the end. Two down, two to go. Now it's time to hang up your lab coat and conical flask, dig out your soldering iron and get stuck into a bit of electronics...

Report: Assembling an Electrical Device

In your day-to-day life you use a vast number of electrical and electronic gadgets — calculators, TVs, computers, electric toenail clippers. All these gadgets are designed and built by <u>electrical engineers</u>.

You Need to Write a Report on Making an Electrical Device

This is the THIRD report for your portfolio for
<u>UNIT 4: USING SCIENTIFIC SKILLS FOR THE BENEFIT OF SOCIETY</u>.

Your report will be in <u>two bits</u>:

1) A <u>DESCRIPTION</u> of your <u>assembled</u> device, including:
   - A <u>description</u> of the <u>function</u> of your device.
   - An <u>explanation</u> of the <u>function</u> of each of its <u>main components</u>.
   - A <u>circuit diagram</u> and <u>photographs</u> of your completed device.

2) <u>TESTING</u> and <u>EVALUATION</u> of your device, including:
   - Tests of your device under <u>conditions of normal use</u>.
   - An <u>evaluation</u> of the performance of your device — does it actually <u>work</u>?
   - Comments on its "<u>fitness for purpose</u>" — is it <u>useful</u> and does it do what it was <u>meant</u> to do?

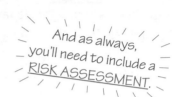
And as always, you'll need to include a <u>RISK ASSESSMENT</u>.

Think Carefully About the Function of Your Circuit

One key to this task is not to get <u>carried away</u>. You're <u>not</u> expected to come up with a fully functional multimedia entertainment system, complete with spinning disco ball and flashing lights.

<u>Keep things simple</u> — there's much less scope for things <u>going wrong</u>. Before you start, think:

1) What <u>COMPONENTS</u> are available? You might need to speak to the technicians to give them time to <u>order</u> any components that aren't in stock.

2) Is the <u>COST</u> reasonable? Don't plan to order loads of components for your circuit before you've checked the costs with your teacher — you might be <u>wasting your time</u>.

3) How much <u>TIME</u> have you got? Building a circuit isn't a 10 minute job — and you need to include time for <u>testing</u> and <u>evaluating</u>. If you run out of time halfway through, you'll <u>lose a lot of marks</u>.

4) What <u>EXACTLY</u> is the <u>function</u> of the device? When you write down its function, be as <u>specific</u> as you can. Then when you do your evaluation, you can check <u>how well</u> it does what it's supposed to do.

<u>YOU MIGHT WANT TO MAKE A CIRCUIT TO:</u>

1) <u>Monitor physical conditions</u> — e.g. monitor the temperature of a machine and light a warning light if it gets too hot.

2) <u>Monitor and control physical conditions</u> — e.g. monitor light levels and switch on a floodlight when it gets dark.

3) <u>Control machines</u> — e.g. automatically open a door when somebody steps on a pressure pad or open a set of gates when your circuit senses the headlights of a car.

The delicate monitoring components in these devices need a very <u>low voltage</u>, but the mechanical components often need a <u>higher voltage</u> supply. An <u>electrical switch</u> called a <u>relay</u> gets round the problem — the low voltage monitoring circuit is used to switch on a <u>separate</u> high voltage circuit (see p.141).

Oh, it has to be USEFUL? — back to the drawing board, then...

All the fancy <u>electronic gadgets</u> in cars would get fried if they had the same current running through them as, say, the starter motor. So relays are used to separate the control circuits. Clever.

Report: Assembling an Electrical Device

You need to select the right <u>components</u> for your circuit, and explain their <u>role</u> in the device.

Electronic Circuits Have Four Main Parts

Most components have one of four <u>basic functions</u> (there are more details on the <u>specific components</u> mentioned further down the page). Your circuit should contain <u>at least one</u>:

1) <u>POWER SOURCE</u> — in your case, a <u>battery</u> or <u>low-voltage power supply</u>. There are strict safety regulations for circuits that run from high-voltage supplies (e.g. mains), so don't plan to build one.

2) <u>INPUT COMPONENT</u> — these act as <u>sensors</u> to detect what's happening in the surroundings, e.g. thermistors, light-dependent resistors, manual switches, variable resistors, etc.

3) <u>PROCESSOR</u> — these <u>take information</u> from the input and decide the <u>output</u>, e.g. transistors.

4) <u>OUTPUT COMPONENT</u> — these give the <u>end result</u> of the device. They usually transfer electrical energy into another form of energy, e.g. a lamp converts electrical energy into light energy.

Circuit Symbols are Used to Represent Components

Each electrical or electronic <u>component</u> has a <u>circuit symbol</u> that's used to represent it in a <u>circuit diagram</u> (see example on the next page). The ones you're most likely to need to use in your report are:

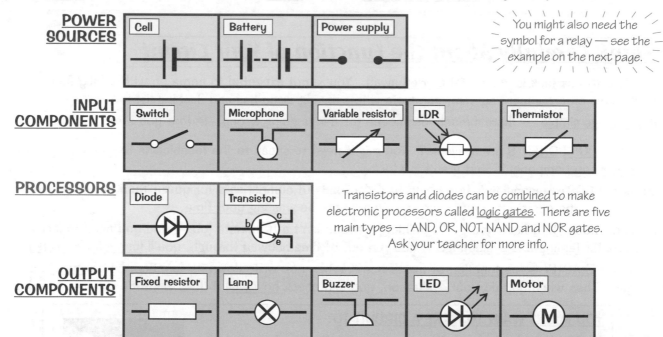

You might also need the symbol for a relay — see the example on the next page.

Transistors and diodes can be <u>combined</u> to make electronic processors called <u>logic gates</u>. There are five main types — AND, OR, NOT, NAND and NOR gates. Ask your teacher for more info.

A <u>VARIABLE RESISTOR</u> is a resistor whose resistance can be <u>changed</u> by twiddling a knob or something.

An <u>LDR</u> (<u>light-dependent resistor</u>) has a <u>high</u> resistance in the <u>dark</u> and a <u>low</u> resistance in <u>bright light</u>.

A <u>THERMISTOR</u> is a resistor with a <u>high</u> resistance when it's <u>cold</u> and a <u>low</u> resistance when it's <u>hot</u>.

A <u>DIODE</u> lets current flow freely through it in <u>one direction</u>, but not in the other.

A <u>TRANSISTOR</u> is basically a <u>posh switch</u>. If the voltage at <u>b</u> is high enough, the transistor switches on and lets current flow through it from <u>c</u> to <u>e</u>.

An <u>LED</u> (<u>light-emitting diode</u>) is a diode (see above) that <u>lights up</u> when a current passes through it.

Input — the kettle on...

You can reduce your <u>risks</u> by using <u>batteries</u> to power your device rather than any of the alternatives — they usually can't supply enough current to give you a serious shock. Avoid mains power like the plague.

Report: Assembling an Electrical Device

Draw a <u>circuit diagram</u> before you start to make your device.

Circuit Diagrams Show How the Components are Connected

<u>EXAMPLE:</u> This circuit is designed to switch on a <u>warning light</u> and a <u>cooling fan</u> when the <u>temperature</u> reaches a set level. The trigger temperature can be set using a <u>variable resistor</u>.

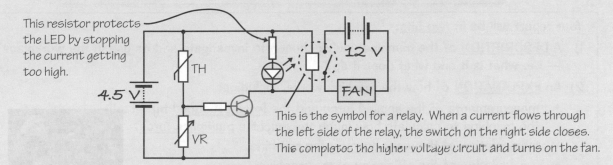

This resistor protects the LED by stopping the current getting too high.

This is the symbol for a relay. When a current flows through the left side of the relay, the switch on the right side closes. This completes the higher voltage circuit and turns on the fan.

1) In this circuit, the 4.5 V from the battery is <u>divided up</u> between the <u>thermistor</u> (<u>TH</u>) and the <u>variable resistor</u> (<u>VR</u>). How the voltage is split depends on the <u>relative resistances</u> of the components — the <u>higher</u> its share of the <u>total resistance</u>, the more of the voltage a component gets.

2) The circuit <u>triggers</u> when the voltage across the <u>VR</u> (and so the voltage across the transistor) is <u>high</u>. That happens when the <u>resistance</u> of the TH is <u>low</u> — think about it.

3) When it's <u>cool</u>, the TH's resistance is <u>high</u> compared with that of the VR, so the transistor isn't triggered and the light and fan are off.

4) But when it's <u>hot</u>, the TH's resistance is <u>low</u> compared with that of the VR. The <u>transistor</u> turns on and completes the rest of the circuit. The <u>LED lights up</u>, the relay switches on the higher voltage circuit and the <u>fan starts</u>.

5) Ta da! Now go back and read that again.

Once you've finished your device, take <u>close-up photos</u> of any bits that you're particularly proud of, and a <u>zoomed-out photo</u> of the whole thing. Then <u>label</u> the photographs and put them in your report.

You Need to Test and Evaluate Your Device

When you've built your device you'll need to <u>test</u> it to make sure it <u>works</u>. It's not enough to just <u>do</u> the tests, though — you have to <u>write</u> about them as you'll need <u>evidence</u> of testing to go in your report.

You also need to test it to make sure it works in <u>exactly</u> the way that you wanted it to. How good is it at doing the job? Does it match your specification?

If you'd built the circuit above, you'd want to know:
1) Does the fan switch <u>on</u> and <u>off</u> at the <u>right temperature</u>?
2) Does the fan run <u>fast enough</u> to give a useful cooling effect?
3) Is the warning light <u>bright enough</u> to be seen?
4) Does the circuit look like it'll <u>last</u> — or does it look ready to fall apart?
5) Is the circuit <u>small enough</u> to make it fit for a practical purpose?

<u>For top marks</u>, you also need to:

Suggest <u>alternative tests</u> for your device.

Suggest <u>improvements</u> to the device to make it <u>more useful</u>.

How do you test a Dalek? E-VAL-U-ATE, E-VAL-U-ATE... (It's the way I tell 'em.)

As you're building your circuit it is a good idea to: 1) Test <u>each component</u> before you fit it to make sure it works. 2) Test <u>each section</u> of your circuit once it's built. That way you won't be trying to find faults once your system is complete — the words "needle" and "haystack" spring to mind.

Report: Using Machines

Simple machines in the workplace (e.g. screwdrivers, spanners, pulleys etc.) act as <u>force multipliers</u> — that is, you need to use <u>less force</u> to do a job using the machine than you would <u>without</u> the machine.

You Need to Write a Report on a Simple Machine

> This is the FOURTH and FINAL (hurrah!) report for your portfolio for <u>UNIT 4: USING SCIENTIFIC SKILLS FOR THE BENEFIT OF SOCIETY</u>.

Your report will be in <u>two bits</u>:

1) A <u>DESCRIPTION</u> of the machine you've chosen to investigate and its <u>use</u> in the workplace — i.e. what is it and what does it do?

2) An <u>EXPLANATION</u> of how the machine works, including:

  • measurements of the <u>applied force</u> and the <u>force produced</u> by the machine to calculate how much the machine <u>multiplies force</u>.
  • a calculation of the <u>work done</u> by the machine.
  • a calculation of the <u>efficiency</u> of the machine.
  • the <u>advantages</u> and <u>disadvantages</u> of <u>friction</u> in machines.

The description's the <u>easy bit</u>, so just a couple of pointers before I move on to the explanation — 1) take a <u>photo</u> of your machine and stick it in your report along with the description (a picture's worth a thousand words), and 2) make sure when you describe its <u>use</u>, it's <u>in the workplace</u> (be specific about it).

Simple Machines All Work in the Same Sort of Way

All simple machines <u>decrease the force</u> you have to use to move something by <u>increasing the distance</u> you need to apply the force over. Some examples of simple machines are:

1) <u>A SINGLE LEVER</u> — e.g. crowbars, spanners, claw hammers.

distance moved with machine
distance moved without machine

2) <u>A DOUBLE LEVER</u> — e.g. pliers, scissors, shears (similar idea to the single lever, but you're pushing from two directions at once).

3) <u>WHEEL AND AXLE</u> — e.g. screwdrivers and wheels for turning things. The outside of the wheel moves much further than the axle.
These are very similar to a spanner.

4) <u>A PULLEY SYSTEM</u> — e.g. 'block and tackle' on a ship. They increase the length of rope you need to pull to move an object.

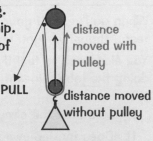

distance moved with pulley
PULL
distance moved without pulley

The amount the machine <u>multiplies the force</u> applied is called the <u>mechanical advantage</u> of the machine. It's a measure of the load (force produced) compared with the effort (force applied).

A <u>high</u> mechanical advantage means you only have to use a <u>small force</u> to move the same load.

$$\text{Mechanical advantage} = \frac{\text{force produced (load)}}{\text{force applied (effort)}}$$

Mechanical advantage — you can fix your own car...

You can use a <u>newton meter</u> to measure the force <u>applied to</u> your machine. How you measure the force <u>produced by</u> the machine depends on its function. For example, if you're using a pulley to lift a load at a steady speed, the force produced by the machine is just the weight (in newtons) of the load.

Report: Using Machines

Calculate the Mechanical Advantage of the Machine

To find the <u>mechanical advantage</u> of your machine, measure the <u>effort</u> and the <u>load</u>, then use the equation.

<u>EXAMPLE</u>: You can use a pulley system to lift a load of 700 N by applying a steady force of 165 N.

$$\text{mechanical advantage} = \frac{\text{force produced}}{\text{force applied}} = \frac{700}{165} = \underline{4.24}$$

Calculate the Work Done by the Machine

The <u>work done</u> by a machine is a measure of its <u>useful energy output</u>.
You find it by multiplying the <u>force produced</u> by the machine by the <u>distance</u> moved by the load.

Work done (J) = Force produced (N) × Distance moved by load (m)

<u>EXAMPLE (cont)</u>: The pulley system lifts the load of 700 N through 12 m.

work done by machine = force produced × distance = 700 × 12 = <u>8400 J</u> (or <u>8.4 kJ</u>)

Machines are Never 100% Efficient

The <u>efficiency</u> of a machine is its useful energy output (<u>work done</u>) as a <u>percentage</u> of the <u>energy input</u>:

$$\text{Efficiency of machine} = \frac{\text{work done by machine}}{\text{energy input}} \times 100\%$$

The <u>energy input</u> is the <u>work done by you</u> as the operator of the machine.

<u>EXAMPLE (cont)</u>: To lift the 700 N load 12 m, you apply a force of 165 N and pull through 60 m of rope.

energy input = work done by operator = force applied × distance = 165 × 60 = <u>9900 J</u> (or <u>9.9 kJ</u>)

$$\text{efficiency of machine} = \frac{\text{work done by machine}}{\text{energy input}} \times 100\% = \frac{8400}{9900} \times 100\% = \underline{84.8\%}$$

Friction is Important in Machines

1) <u>Friction</u> is a force that <u>opposes movement</u>.
2) When you want surfaces to <u>slide across each other</u>, e.g. when pushing a box up a ramp or turning a wheel on an axle — friction is a <u>nuisance</u>. It makes the surfaces heat up, which wastes energy.
3) It's because some of the energy input is wasted as heat that machines <u>aren't 100% efficient</u> — you need to put <u>more energy</u> into the machine than you get out to <u>overcome</u> the friction.
4) To <u>reduce</u> friction and improve the efficiency of machines you can use <u>lubricants</u>, like oil.
5) Friction can sometimes be <u>useful</u> though, e.g. it prevents heavy loads from <u>slipping</u> down slopes.
6) It's also worth remembering that machines <u>wouldn't work at all</u> if there was <u>no friction</u>. E.g. in the pulley system, it's only the friction between the <u>rope</u> and your <u>hand</u> that lets you pull it at all.

Nearly there now...

Right — here we go again. All together now... <u>don't forget to do a risk assessment</u>.
And do it properly — even if you <u>do</u> feel like a muppet writing a risk assessment for a spanner.

Tips on Producing Your Portfolios

Even if you think this stuff is <u>blindingly</u> obvious, <u>READ IT</u> anyway — humour me.
It's a list of the stuff you <u>must</u> remember when you're putting your portfolios together...

You'll Need a Portfolio for Each Coursework Unit

1) You'll have to produce <u>three separate portfolios</u> — one each for <u>Unit 1</u>, <u>Unit 3</u> and <u>Unit 4</u>.

2) For each unit, you'll have to write various <u>reports</u> (see pages 5, 11, 131 and 132-143 for specific advice on the reports).

3) The portfolios are marked by your <u>teacher</u> and moderated by AQA.

4) The portfolios make up <u>two thirds</u> of your final grade, so they're pretty important...

I'm not impressed

Your Portfolios Should be Neat and Easy to Follow

If you hand in a <u>jumbled</u>, <u>illegible mess</u> and call it a portfolio, your teacher will <u>NOT</u> be impressed.

1) Your portfolios should be <u>well organised</u>, <u>well structured</u> and <u>tailored</u> to the tasks (so no random notes from lessons, no unidentified graphs or diagrams, no pictures of Elvis).

2) If you've got access to a computer, <u>word process</u> your reports — they're much <u>neater</u> that way, and it's easier to <u>edit</u> your work if you change your mind about something.

3) Make life easy for your marker — break up your report with <u>headings</u> to make it easier to follow.

4) If you're including any <u>graphs</u>, <u>diagrams</u> or <u>photos</u>, make sure they're clearly <u>labelled</u>.

5) There's no right or wrong <u>length</u> for a report. But they should be only as long as they <u>need to be</u> to cover everything. Don't <u>pad them out</u> for the sake of it — no one likes wading through waffle.

6) <u>Read through</u> your work carefully before handing it in (run a <u>spellcheck</u> if you're using a computer).

Make Sure It's All Your Own Work

Make sure there's nobody else's work in with yours. <u>I</u> know you're honest, but AQA take a very dim view of two candidates' work being <u>too similar</u>.

It's fine to include bits in your reports that come from <u>books</u> or <u>websites</u>, but you need to <u>reference</u> them — say where they come from. Your references can go at the <u>end</u> of the report.

You also need to work as <u>independently</u> as possible. The more <u>help</u> you need from your <u>teacher</u>, the lower your mark. But, saying that, it's better to do something with help than just miss it out altogether.

And Then for a Few Finishing Touches

Clear presentation makes your portfolio <u>easier to follow</u>... which makes life easier for the person <u>marking</u> it... which puts them in a <u>good mood</u>... which has got to be good. Here are a <u>few tricks</u>:

1) Make a <u>front cover</u> for your portfolio. It should have <u>your name</u>, the <u>course name</u> and the <u>unit number and title</u>. (There's an official cover sheet to go in front of this as well — ask your teacher.)

2) Separate the different reports with <u>header pages</u> — nothing fancy, just put the name of the report.

3) <u>Number</u> your pages. Call the first header page "page 1", then just <u>number through</u> to the end.

4) Include a <u>contents page</u> with page numbers.

5) Hole-punch everything and put it in a <u>ring binder</u>... and you're done. Woohoo!

Index

Index

Index

Index and Answers

Answers

Revision Summary for Section 2.2 (page 28)

12) a) no

b) yes

c) no

Revision Summary for Section 2.6 (page 72)

5) 29

6) 6

9) 4

10) 2

12) a) $2Mg + O_2 \rightarrow 2MgO$

b) $Cl_2 + 2KBr \rightarrow Br_2 + 2KCl$

c) $C_6H_{12}O_6 \rightarrow 2C_2H_5OH + 2CO_2$

d) $2Na + 2H_2O \rightarrow 2NaOH + H_2$

Revision Summary for Section 2.7 (page 81)

31) a) metal

b) laminated glass or reinforced glass

c) ceramic, e.g. porcelain

d) ceramic or heat-resistant polymer

Bottom of page 86

Energy = Power × Time = 0.06 kW × 3 h = <u>0.18 kWh</u>

Bottom of page 88

TV: (7 ÷ 200) × 100 = 3.5%

Loudspeaker: (0.5 ÷ 35) × 100 =1.4%

Revision Summary for Section 2.8 (page 94)

6) Current = Power ÷ Voltage = 1500 W ÷ 230 V
= 6.52 A so a <u>7 A fuse</u>

10) Energy = Power × Time = 550 W × 30 s
= <u>16 500 J</u> (16.5 kJ)

11) Energy. Energy = Power × Time so
2 kWh = 2000 W × 3600 s = <u>7 200 000 J</u>
(= 7200 kJ / 7.2 MJ)

12) 1000 W = 1 kW so 100 W = 0.1 kW so:

a) 0.1 kW × 3 h = <u>0.3 kWh</u>

b) 0.1 kW × 0.5 h = <u>0.05 kWh</u>

13) Energy used = 0.2 kW × 0.25 h = 0.05 kWh
Cost = 0.05 × 11 = <u>0.55p</u>

15) a) (40 ÷ 100) × 100 = 40%

b) 100 J – 40 J = 60 J

16) Useful output = 1000 J – 230 J = 770 J
Efficiency = (770 J ÷ 1000 J) × 100% = <u>77%</u>

Revision Summary for Section 2.9 (page 102)

1) Speed = Distance ÷ Time = 6 ÷ 2 = 3 m/s

2) Distance = Speed × Time

a) 18 × 2 = 36 m

b) 1 minute = 60 s. 18 × 60 = 1080 m (= 1.08 km)

3) Time = Distance ÷ Speed = 500 ÷ 2 = 250 s
(= 4 min 10 s)

5) Acceleration = Change in Speed ÷ Time = (12 – 0) ÷ 5
= 2.4 m/s^2

6) Deceleration = 18 ÷ 6 = 3 m/s^2

7) Deceleration = (7 – 1.5) ÷ 2 = 5.5 ÷ 2 = 2.75 m/s^2

10) Distance = Speed × Time = 15 × 0.5 = 7.5 m